Biotherapeutics: mAbs, ADCs and Factors

BIOTHERAPEUTICS: MABS, ADCS AND FACTORS

Shivsanjeevi Sripathi

Delve PUBLISHING

www.delvepublishing.com

Biotherapeutics: mAbs, ADCs and Factors

Shivsanjeevi Sripathi

Delve Publishing
224 Shoreacres Road
Burlington, ON L7L 2H2
Canada
www.delvepublishing.com
Email: orders@arclereducation.com

ISBN: 978-1-77469-375-9 (Hardcover)

ABOUT THE AUTHOR

ShivSanjeevi Sripathi completed his Masters in Biotechnology from Mumbai University in 2008. He was awarded for academic excellence in both his Bachelors and Masters for securing second rank in Mumbai University in 2006 and first rank in his college: Kishinchand chellaram College. For his Masters he secured first rank in his college KET's V.G.Vaze College. He qualified CSIR and NET and and TOEFL in September 2008. He then worked on a stem cell project at the Specialized Centre for Cell Based Therapy (SCCT),KEM Hospital at Mumbai on a project entitled, : Isolation & detection of stem cells from Human Umbilical cord/ amniotic membrane" following which he worked at Junior Research Fellow at Microbiology & Cell Biology Department, Indian Institute of Science, Bangalore on cloning of cell wall genes and transcription factors in E.coli & M.smegmatis. As a writer, he has authored and co-authored 35 books on various aspects of biology such as bionics, molecular wires, cloning, hypertension, the epidemics of the 21st century, handling depression, camouflage, hygiene, immunology and many more with international publishers. He loves to read and share on interesting aspects of life sciences in books. In his free time he loves to travel and explore and give talks on spirituality, ancient customs and traditions.

TABLE OF CONTENTS

PREFACE

Biotherapeutics refer to therapeutic approaches with the active substance extracted or produced from a biological source. This encompasses monoclonal antibodies (mAbs), stem cell therapies, gene therapy, growth factors and cytokines, to name a few.

"We are at the beginning of a new era of immunochemistry, namely the production of "antibody based" molecules"

— César Milstein: From Nobel Lecture (8 Dec 1984), collected in Tore Frängsmyr and Jan Lindsten (eds.), Nobel Lectures in Physiology Or Medicine: 1981-1990 (1993)

This book on biotherapeutics comes at a time when the market for these products is on the rise. For instance, of 46 drugs that were approved by the United States Food and Drug Administration (USFDA) in 2017, ten were mAbs (Johnson D. E, *International journal of molecular sciences*, 2018). Further, despite 2020 being "the year of COVID-19", 10 of the 13 biologics were mAbs that gained FDA approval (Torre and Albericio, *Molecules*,2021). One can comprehend the potential of these molecules called "magic bullets" given that Casirivimab and Imdevimab (Regeneron Pharmaceuticals Inc) received emergency use authorization (EUA) from the FDA for treatment of mild to moderate COVID-19 infection in adults and pediatric patients.

In addition to the direct targeting of a slew of maladies by mAbs via ADCC or CDC or cell death induction, there are "additions" to mAbs in the form of bispecific antibodies, antibody-drug conjugates (ADCs), chimeric antigen receptor T cell (CAR-T) therapy and antibody-radionuclide conjugates. The employment of high-throughput sequencing, robotic screening, novel biomarkers, combination with other therapeutic systems such as radiotherapy can all augment mAb-based therapy to realize the goals of biotechnology viz. optimal health for all.

The book apart from discussing mAbs available for cancer, also highlights the research done for mAb therapy for asthma and COVID-19. An example is LY-CoV555 reported by Bryan E. Jones and team (2021) in Science Translational Medicine that lowered SARS-CoV replication and load. There are also recent developments discussed such as of "glycoengineering" of mAbs aimed at altering the sugar residues to augment desired properties. Further, a detailed discussion is presented on the fastest-growing drugs called antibody–drug conjugates (ADC) composed of a mAb conjugated to a cytotoxic payload via a chemical linker targeting a specific antigen thus augmenting the specificity of treatment. As of now, nine ADCs have received FDA approval for two solid tumors and six hematological malignancies. ADCs have entered the clinical field for treatment and studies on these "bullets" are ongoing. The detailed scrutiny of linkers, payloads and conjugation approaches can further augment the development of more ADCs to enter the market to quell not only cancer but also other diseases. The impact and research on cytokines and growth factors as therapeutics are also discussed in this book. The concept of nanobodies is also presented.

Conventional drug design entails the targeting of active sites of proteins: the field of nucleic acid therapeutics looks at circumventing the challenges of correct folding and post-translational modifications warranted by the former. The FDA approved a DNA therapeutic, Zolgensma in 2019 that is aimed at spinal muscular atrophy (SMA) by replacing the function of the non-functional SMN1 gene with a functional SMN gene. There are also exciting clustered regularly interspaced short palindromic repeats and CRISPR-associated protein 9 (CRISPR-Cas9)-based systems in the pipeline. An example is CTX001 from CRISPR Therapeutics that disrupts the BCL11A gene in stem cells to boost γ-globin and serve as an alternative to a patient's defective adult hemoglobin (HbA) seen in sickle cell disease and beta-thalassemia is undergoing trials.

The promise of RNA therapeutics is highlighted by the ongoing COVID-19 pandemic where in addition to a slew of molecules being scrutinized as vaccines, the first two of all COVID-19 vaccines under development to have received emergency use authorization by the FDA were RNA-based: BNT162b1 and BNT162b2. Emergency use authorization was later also issued to Moderna RNA vaccine: its time-line of development is interesting: the appropriate sequence for the Moderna vaccine candidate sequence was selected 2 days after the SARS-CoV-2 genome sequence was deciphered, the first clinical batch of mRNA-1273 was produced after 25 days and dosing the first participant was done following another 35 days. Such short-time frames over conventional systems have elicited much interest given the epidemic outbreaks attributed to newly emerging viruses.

This book is one of the first in compiling select biotherapeutics-based systems viz. mAbs, ADCs, cytokines, growth factors and nucleic acids. This paves the way for part 2 encompassing the remaining technologies with research-backed data. This research-backed book presents insights and developments for several maladies that plague humanity including case studies of COVID-19.

It is dedicated to each of you who has overcome.

"The science of today is the technology of tomorrow"

-Edward Teller

CHAPTER 1

Biotherapeutics-a Snapshot

Contents

1.1 INTRODUCTION

Biotherapeutics or biologicals are molecules aimed at targeting maladies with the active substance extracted or produced from a biological source. These encompass stem cell-based approaches, gene editing, vaccines, growth factors, cytokines, gene silencing, cell-based products, monoclonal antibodies (mAbs), tissue engineering and recombinant proteins (Johnson, 2018).

Drug delivery entails the carrying of a therapeutic molecule to a particular location and its release at an optimal rate. These systems can deliver a slew of "biotherapeutics" ranging from nucleic acids, peptides and other molecules. The shape, surface charge, size along with stability facilitate the categorization of these molecules. Points that are essential to be assimilated are the possible inflammation and adherence to requisite regulations and large-scale production. The delivery systems suitable for biotherapeutics are illustrated in figure 1 below (Mohammad A. Obeid et al, 2017: *This article is distributed under the terms of the Creative Commons Attribution Licence 3.0 which permits any use, distribution, and reproduction in any medium, provided the original author(s) and the source are credited. To view a copy of the license, visit http://creativecommons.org/licenses/by/3.0/*):

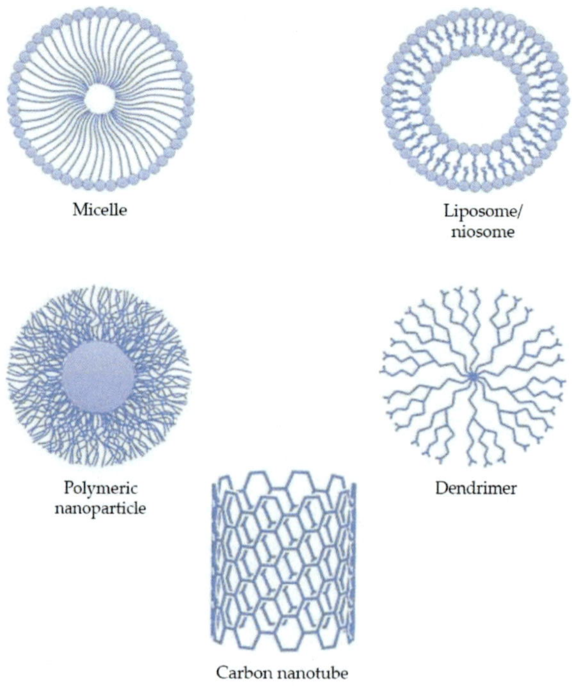

A slew of natural products have been scrutinized to target several maladies employing delivery systems as summarized in table 1 below (Mohammad A. Obeid et al, 2017: *This article is distributed under the terms of the Creative Commons Attribution Licence 3.0 which permits any use, distribution, and reproduction in any medium, provided the original author(s) and the source are credited. To view a copy of the license, visit http://creativecommons.org/licenses/by/3.0/*):

Plant/constituents	**Biological activity**	**Delivery system used**	**Efficacy of the delivery system**
Curcumin	Anticancer	Liposomes	Long systemic residence time and high entrapment efficiency
Curcumin	Anticancer and anti-oxidant	Phytosomes	For enhanced antioxidant action and bioavailability
Curcumin	Anticancer	Emulsion system	To improve absorption
Curcumin	Anticancer and anti-oxidant	Transferosomes	To enhance permeation
Curcumin	Anti-inflammatory	Micropellatization	To enable sustained release and for specific locus targeting
Quercetin	Anticongestion and antianxiety	Liposomes	To increase efficacy, bioavailability and reduce side effects
Quercetin	Anti-inflammatory and antioxidant	Microspheres	To enhance permeation
Quercetin	Antioxidant	Emulsion system	To enhance permeation
Ginkgo biloba	Brain activator	Nanoparticles	To improve cerebral blood flow and metabolism
Ginkgo biloba	Antiasthmatic, antidiabetic and cardioprotective	Phytosomes	To improve efficacy
Wogonin	Anticancer	Liposomes	To increase duration of action
Embelin	Antifertility and anti-bacterial	Phytosomes	To enhance solubility

The flexibility of antibodies and their ability to recognize a slew of molecules reflect on the initial approval of 5-8 antibody-based approaches in 2014 to 53 in Phase III studies in 2016 (Mohammad A. Obeid et al, 2017). These can be murine (the suffix "-momab" is attached), chimeric, humanized, and fully human. An illustration of monoclonal therapeutic antibodies (MTAs) is given in figure 2 below: ***TOP***: An antibody; ***BOTTOM***: MTAs with green-human origin; yellow-murine origin (Monireh Golpour et al, 2021: *Copyright © 2021 Monireh Golpour et al. This is an open access article distributed under the Creative Commons Attribution License, which permits unrestricted use, distribution, and reproduction in any medium, provided the original work is properly cited*):

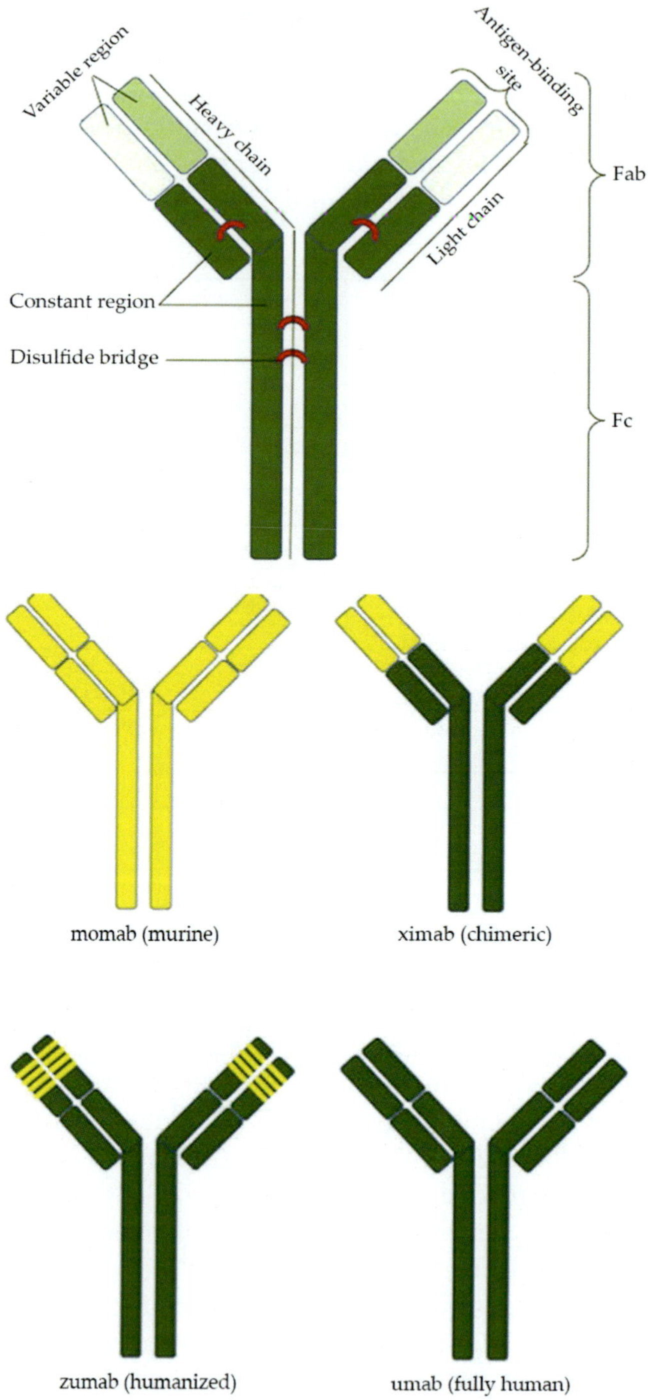
Variable region
Heavy chain
Antigen-binding site
Fab
Light chain
Constant region
Disulfide bridge
Fc
momab (murine)
ximab (chimeric)
zumab (humanized)
umab (fully human)

For instance, analyses of MTAs by Monireh Golpour et al (2021) between 2006 and 2018 based on data from the Food and Drug Deputy of Mazandaran University of Medical Sciences revealed the predominant use of MTAs for cancer and a "dramatic" rise in the types and uses of MTAs (124.3-fold increase). The antibodies scrutinized and the rise in sales are shown in table 2 and figure 3 below (ONCO-oncology; AIID: arthritis, inflammation and immune disorder (Monireh Golpour et al, 2021: *Copyright © 2021 Monireh Golpour et al. This is an open access article distributed under the Creative Commons Attribution License, which permits unrestricted use, distribution, and reproduction in any medium, provided the original work is properly cited*):

Generic name	Isotype	Target	Indications	Therapy field	FDA approval date	Source
Gemtuzumab	IgG4	CD33	Acute myeloid leukemia	ONCO	2000	Humanized
Cetuximab	IgG1K	EGFR	Head and neck cancer, KRAS-ve colorectal cancer	ONCO	2012	Chimeric
Infliximab	IgG1K	TNF-α	Crohn's disease, ulcerative colitis, RA, ankylosing spondylitis, psoriatic arthritis	AIID	1998	Chimeric
Trastuzumab	IgG1K	HER2	Breast cancer, gastric/gastroesophageal carcinoma	ONCO	1998	Humanized
Bevacizumab	IgG1K	VEGF-A	Colorectal cancer, nonsquamous non-small-cell lung cancer, renal cell carcinoma, glioblastoma	ONCO	2004	Humanized

Rituximab	IgG1K	CD20	Non-Hodgkin's lymphoma, chronic lymphocytic leukemia, rheumatoid arthritis	ONCO, AIID	1997	Chimeric
Adalimumab	IgG1K	TNF-α	Rheumatoid arthritis, juvenile idiopathic arthritis, psoriatic arthritis, ankylosing spondylitis, Crohn's disease, plaque psoriasis	AIID	2002	Fully human
Natalizumab	IgG4K	Alpha-4 integrin	Multiple sclerosis, Crohn's disease	AIID	2004	Humanized
Daclizumab	IgG1	CD25	Acute transplant rejection, multiple sclerosis	AIID	1997	Humanized
Denosumab	IgG2	RANK ligand	Bone metastases from solid tumors, osteoporosis, giant cell tumor of the bone	ONCO	2010	Fully human
Belimumab	IgG1	BAFF, BLYs	Systemic lupus erythematosus	AIID	2011	Fully human
Tocilizumab	IgG1k	IL-6R	Rheumatoid arthritis	AIID	2010	Humanized
Ranibizumab	IgG1k	VEGF-A	Age-related macular degeneration, age-related vision loss	AIID	2015	Humanized
Abciximab	IgG1	CD41	Acute myocardial infarction	AIID	1993	Chimeric

Pertuzumab	IgG1k	HER2	Metastatic HER2-positive breast cancer, neoadjuvant in early HER2-positive breast cancer	ONCO	2012	Humanized
Omalizumab	IgG1k	IgE FC region	Severe allergic asthma	AIID	2014	Humanized
Alemtuzumab	IgG1k	CD52	Chronic lymphocytic leukemia (CLL), cutaneous T-cell lymphoma (CTCL), multiple sclerosis	ONCO, AIID	2013	Humanized

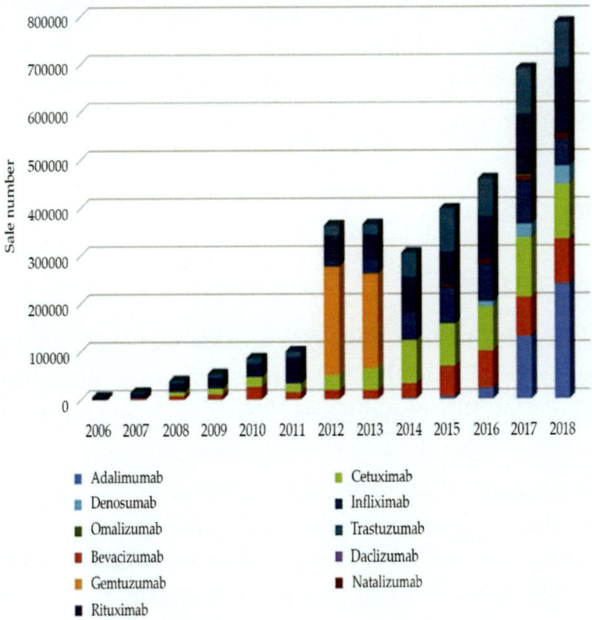

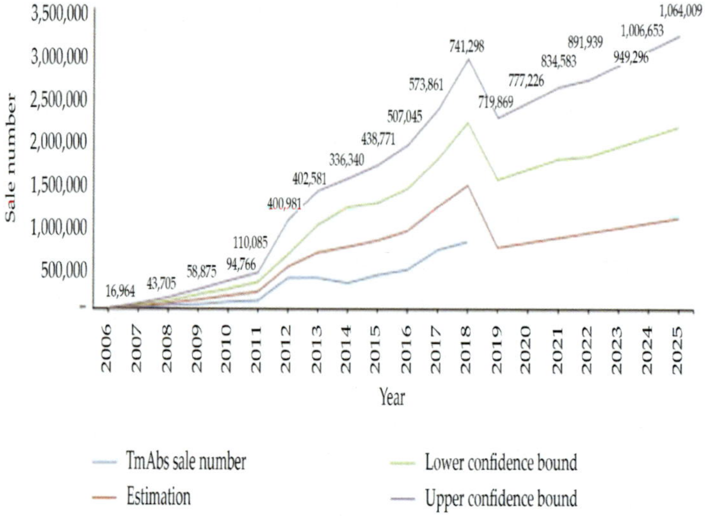

Amalgamating the properties of antibodies with small molecule drugs specifically targeting malignancies brings in the field of antibody–drug conjugates (ADCs) that circumvents off-target toxicity. The first-generation ADCs were withdrawn on account of toxicity and low potency. This led to second-generation ADCs: Acetris® (brentuximab–vedotin) and Kadcyla® (trastuzumab–emtansine). Sharpening of limitations such as resistance and inability to target solid tumors has led to third-generation ADCs under development (Mohammad Obeid et al, 2017).

1.2 TRENDS

1.2.1 The Year 2017

According to research in *Molecules (Basel, Switzerland)* by de la Torre and Albericio, the year 2017 saw the FDA approving 12 biologics and 34 New Chemical Entities (NCEs) considered *"a record in terms of new entities approved by the FDA"* over the last two decades. Figure 4 below is illustrative of this trend (Copyright © 2018 by the authors: de la Torre and Albericio, 2017: Licensee MDPI, Basel, Switzerland. *This article is an open access article distributed under the terms and conditions of the Creative Commons Attribution (CC BY) license: http://creativecommons.org/licenses/by/4.0/*):

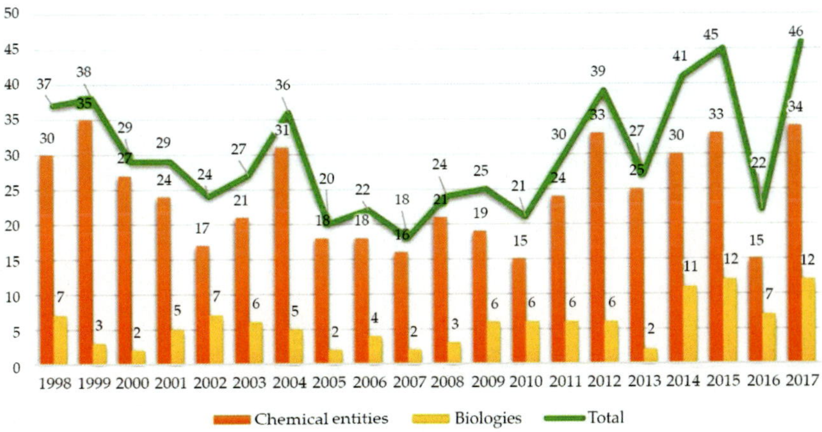

Of these 12 biologics, one is an antibody-drug conjugate (ADC), nine are antibodies and two are enzymes. This is depicted in table 3 below (Copyright © 2018 by the authors: de la Torre and Albericio, 2017: Licensee MDPI, Basel, Switzerland. *This article is an open access article distributed under the terms and conditions of the Creative Commons Attribution (CC BY) license: http://creativecommons.org/licenses/by/4.0/*):

Key: [a] USA; [b] antibody; [c] ADC; [d] enzyme

Active Ingredient	Trade Name [a]	Disease
Avelumab	Bavencio™ [b]	Merkel cell carcinoma
Dupilumab	Dupixent™ [b]	Asthma
Benralizumab	Fasenra™ [b]	Asthma
Emicizumab	Hemlibra™ [b]	Hemophilia A
Durvalumab	Imfinzi™ [b]	Urothelial carcinoma
Sarilumab	Kevzara™ [b]	Rheumatoid arthritis
Ocrelizumab	Ocrevus™ [b]	Multiple sclerosis
Brodaluma	Siliq™ [b]	Psoriasis
Guselkumab	Tremfya™ [b]	Psoriasis
Inotuzumabozogamicin	Besponsa™ [c]	Acute lymphoblastic leukemia
Cerliponasealfa	Dupixent™ [d]	Batten disease
Vestronidasealfa	Fasenra™ [d]	Sly syndrome

Among these, natural products inspired the molecules: a noroxymorphone-derivative Naldemedine (Symproic™) for constipation induced by opoids; Deflazacort (Emflaza™), a glucocorticoid is for Duchenne muscular dystrophy and ertugliflozin (Steglatro™) with glucose is for type 2 diabetes. The structures of molecules are illustrated in figure 5 below (Copyright © 2018 by the authors: de la Torre and Albericio, 2017: Licensee MDPI, Basel, Switzerland. *This article is an open access article distributed under the terms and conditions of the Creative Commons Attribution (CC BY) license: http://creativecommons.org/licenses/by/4.0/*):

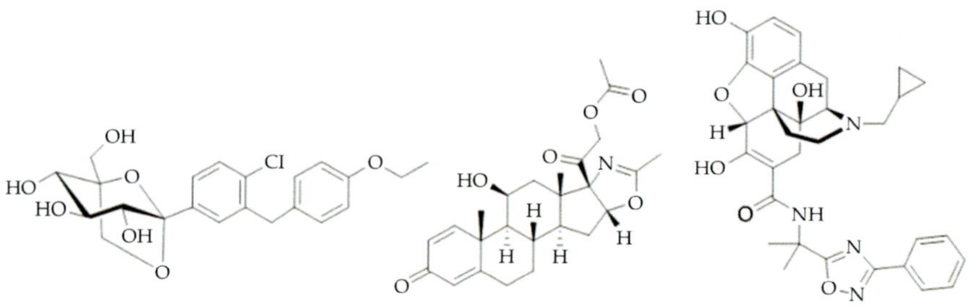

1.2.2 The Year 2018

This year saw the approval of 59 new drugs (42 New Chemical Entities (NCE) and 17 Biologics) by the FDA termed as a "record" (G de la Torre and Albericio, 2019). Figure 6 below illustrates the pattern (Copyright © 2019 by the authors: G de la Torre and Albericio, 2019; Licensee MDPI, Basel, Switzerland. *This article is an open access article distributed under the terms and conditions of the Creative Commons Attribution (CC BY) license:http://creativecommons.org/licenses/by/4.0/*):

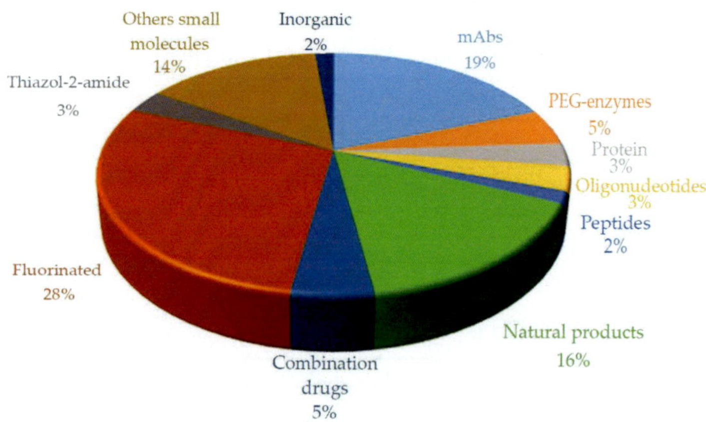

The biologics are presented below in table 4 (Copyright © 2019 by the authors: G de la Torre and Albericio, 2019; Licensee MDPI, Basel, Switzerland. *This article is an open access article distributed under the terms*

and conditions of the Creative Commons Attribution (CC BY) license:http:// creativecommons.org/licenses/by/4.0/):

Active Ingredient (in alphabetical order)	Trade Name (USA)	Class	Disease
Burosumab	Crysvita™	monoclonal antibody	X-linked dominant hypophosphatemic rickets
Calaspargasepegol	Asparlas™	pegylated enzyme	Acute lymphoblastic leukemia
Cemiplimab	Libtayo™	monoclonal antibody	Cutaneous squamous cell carcinoma
Cenegermin	Oxervate™	protein	Neurotrophic keratitis
Elapegademase	Revcovi™	pegylated enzyme	Adenosine deaminase severe combined immunodeficiency
Emapalumab	Gamifant™	monoclonal antibody	Hemophagocyticlymphohistiocytosis
Erenumab	Aimovig™	monoclonal antibody	Migraine prevention
Fremanezumab	Ajovy™	monoclonal antibody	Migraine prevention
Galcanezumab	Emgality™	monoclonal antibody	Migraine prevention
Ibalizumab	Trogarzo™	monoclonal antibody	Multidrug-resistant HIV-1
Lanadelumab	Takhzyro™	monoclonal antibody	Hereditary angioedema attacks
Mogamulizumab	Poteligeo™	monoclonal antibody	Relapsed or refractory mycosis fungoides and Sézary disease
Moxetumomabpasudotox	Lumoxiti™	monoclonal antibody	Relapsed or refractory hairy cell leukemia
Pegvaliase	Palynziq™	pegylated enzyme	Phenylketonuria
Ravulizumab	Ultomiris™	monoclonal antibody	Paroxysmal nocturnal hemoglobinuria and atypical hemolytic uremic syndrome
Tagraxofusp-erzs	Elzonris™	fusion protein	Blasticplasmacytoid dendritic cell neoplasm
Tildrakizumab	Ilumya™	monoclonal antibody	Moderate-to-severe plaque psoriasis

1.2.3 The Year 2019

Among the monoclonal antibodies (mAbs) approved, is Caplacizumab (Cablivi™) is the first "nanobody" in the market. Table 5 documents the list of these molecules (de la Torre and Albericio, 2020: Copyright © 2020 by the authors: Licensee MDPI, Basel, Switzerland. *This article is an open access article distributed under the terms and conditions of the Creative Commons Attribution (CC BY) license: http://creativecommons.org/licenses/by/4.0/*):

Active Ingredient (in alphabetical order)	**Trade Name (USA)**	**Class**	**Indication**
Brolucizumab	Beovu™	Monoclonal antibody	Wet age-related macular degeneration
Caplacizumab	Cablivi™	Monoclonal antibody (Nanobody)	Thrombosis and thrombotic thrombocytopenic purpura
Crizanlizumab	Adakveo™	Monoclonal antibody	Prevention of vaso-occlusive crisis in patients with Sickle cell anemia
Risankizumab	Skyrizi™	Monoclonal antibody	Moderate to severe plaque psoriasis
Romosozumab	Evenity™	Monoclonal antibody	Osteoporosis in postmenopausal women
Enfortumabvedotin-ejfv	Padcev™	Antibody drug conjugate	Cancers expressing Nectin-4
Polatuzumabvedotin- piiq	Polivy™	Antibody drug conjugate	Diffuse large B-cell lymphoma
Fam-trastuzumabderuxtecan-nxki	Enhertu™	Antibody drug conjugate	Unresectable or metastatic HER2-positive breast cancer
Luspatercept	Reblozyl™	Fusion Protein	Anemia in β-thalassemia and myelodysplastic syndromes
Prabotulinumtoxin (botulinum toxin)	Jeuveau™	Neurotoxin protein	Temporary improvement of frown lines

Natural products were the inspiration for a slew of drugs that received approval from the FDA. Cefiderocol (Fetroja™) of the cephalosporin family is indicated for multidrug-resistant Gram-negative bacteria-caused urinary tract infection. The steroid allopregnanolone or brexanolone (Zulresso™) is for depression; nucleic acid base-derived Istradefylline (Nourianz™) and Fedratinib (Inrebic™) are for Parkinson's disease and myeloproliferative disease therapy, respectively. Solriamfetol (Sunosi™) from phenylalanine is for sleepiness; lefamulin (Xenleta™) is for community-acquired bacterial pneumonia and fluorodopa (FDOPA™), the fluorine-18 isotopologue of L-DOPA is for visualizing nerve cells of Parkinson's disease patients by PET. These are shown below in figure 7 (de la Torre and Albericio, 2020: Copyright © 2020 by the authors: Licensee MDPI, Basel, Switzerland. *This article is an open access article distributed under the terms and conditions of the Creative Commons Attribution (CC BY) license: http://creativecommons.org/licenses/by/4.0/*):

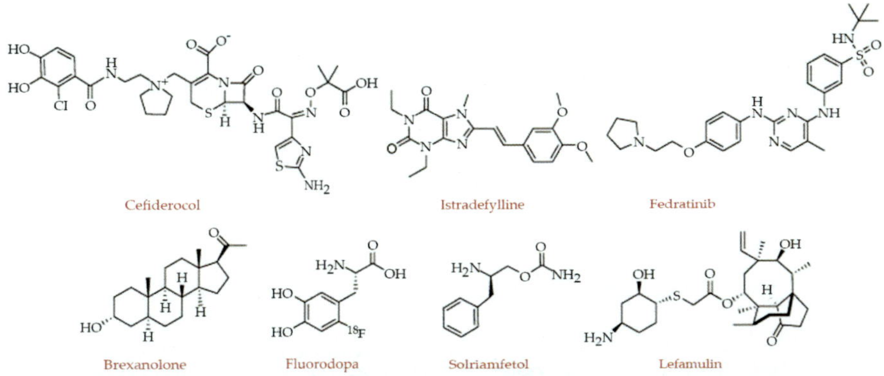

A marine mollusk-derived molecule called monomethylauristatin E [MMAE: brentuximabvedotin (Adcetris™)] received approval as two antibody-drug conjugates (ADCs): enfortumabvedotin-ejfv (Padcev™) and polatuzumabvedotin-piiq (Polivy™) containing the linker Val-Cit with the auto-inmolative moiety p-aminobenzyl alcohol carbamate (PABC). The other ADC was Fam-trastuzumabderuxtecan-nxki (Enhertu™) with the tetrapeptide Gly–Gly–Phe–Gly (GGFG) linker. All these ADCs are for tumors. The structures of these molecules are depicted below in figure 8: ***TOP***: enfortumabvedotin and polatuzumabvedotin; ***BOTTOM***: Fam-trastuzumabderuxtecan(de la Torre and Albericio, 2020: Copyright © 2020 by the authors: Licensee MDPI, Basel, Switzerland. *This article is an open access article distributed under the terms and conditions of the Creative Commons Attribution (CC BY) license: http://creativecommons.org/licenses/by/4.0/*):

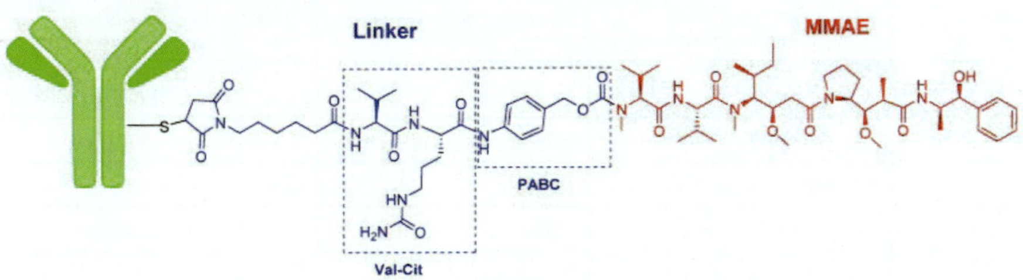

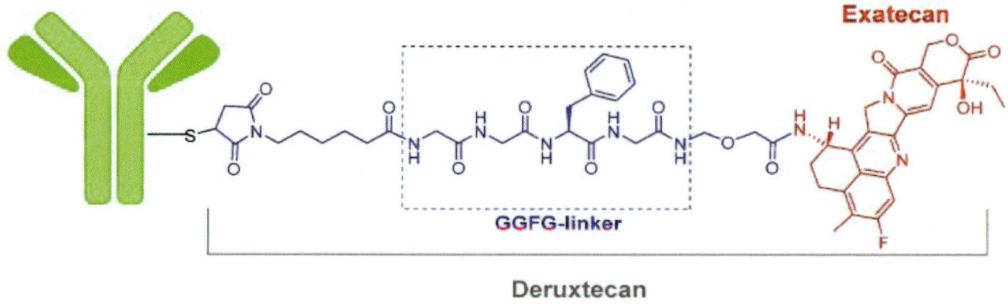

1.2.4 The Year 2020

Notwithstanding COVID-19, 13 biologics and 40 New Chemical Entities (NCEs) received FDA approval. 10 of the 13 biologics were mAbs as presented below in table 6 (Torre and Albericio, 2021: Copyright © 2021 by the authors; Licensee MDPI, Basel, Switzerland. *This article is an open access article distributed under the terms and conditions of the Creative Commons Attribution (CC BY) license: http://creativecommons.org/licenses/by/4.0/*):

Trade name (USA)	Active Ingredient	Class	Indication
Blenrep™	Belantamabmafodotin-blm	Antibody-drug conjugate	Multiple myeloma
Danyelza™	Naxitamab-gqgk	Monoclonal antibody	High-risk refractory or relapsed neuroblastoma
Ebanga™	Ansuvimab-zykl	Monoclonal antibody	Ebola virus
Enspryng™	Satralizumab-mwge	Monoclonal antibody	Neuromyelitisoptica spectrum disorder
Inmazeb™	Atoltivimab, Aaftivimab, Odesivimab-ebgn	Monoclonal antibodies	Ebola virus
Margenza™	Margetuximab	Monoclonal antibody	HER2+ breast cancer
Monjuvi™	Tafasitamab-cxix	Monoclonal antibody	Relapsed or refractory diffuse large B-cell lymphoma
Sarclisa™	Isatuximab	Monoclonal antibody	Multiple myeloma
Sogroya™	Somapacitan-beco	Protein	Growth hormone
Tepezza™	Teprotumumab-trbw	Monoclonal antibody	Thyroid eye disease
Trodelvy™	Sacituzumabgovitecan-hziy	Antibody-drug conjugate	Metastatic triple-negative breast cancer
Uplizna™	Inebilizumab-cdon	Monoclonal antibody	Neuromyelitisoptica spectrum disorder
Vyepti™	Eptinezumab-jjmr	Monoclonal antibody	Preventive treatment of migraine in adults

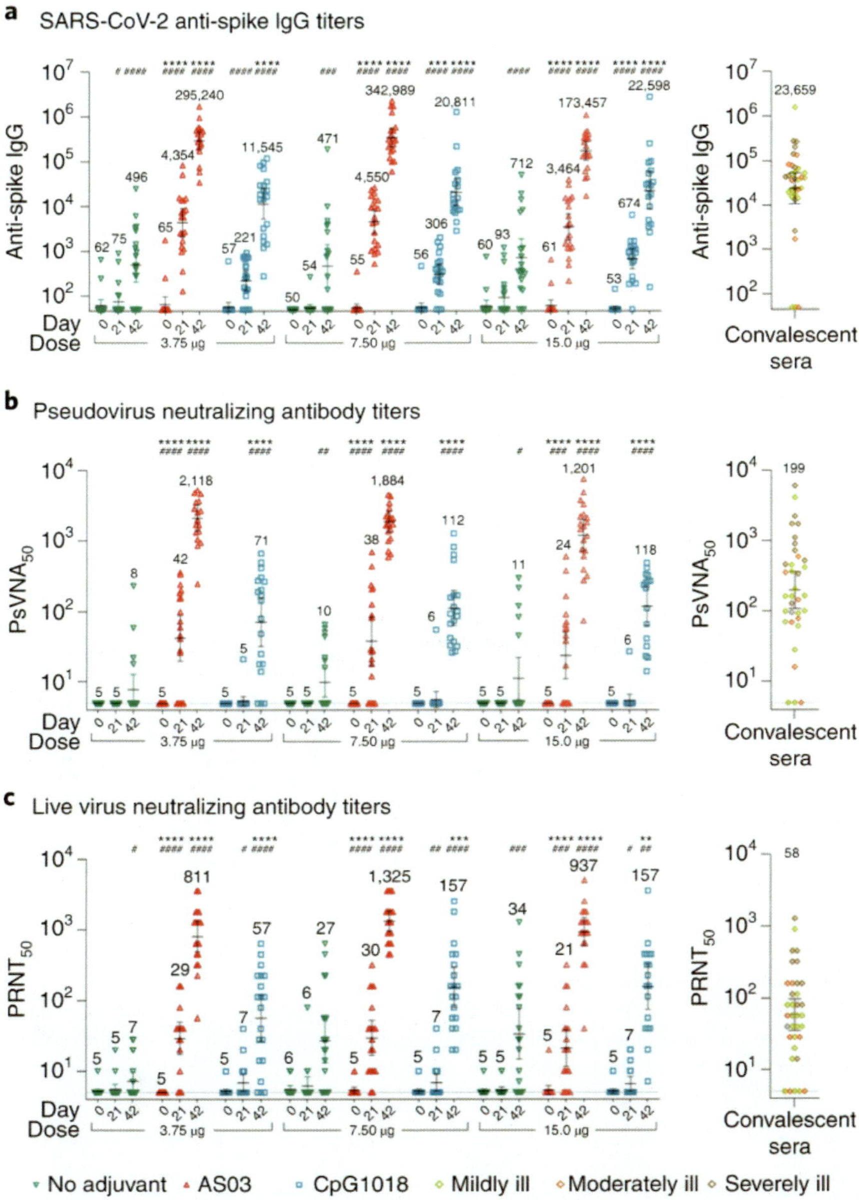

In terms of T-cells, interferon (IFN)-γ and interleukin (IL)-4 responses in peripheral blood mononuclear cells (PBMCs) were significant in both presence and absence of the adjuvant. Figure 13 below illustrates the responses of IFN-γ (**a**) or IL-4 (**b**) at baseline (Day 0) and 21 d after the first (Day 21) or second (Day 42) following the vaccine: SPC, spot-forming cell. Significant differences in treatment are depicted as $^{\#}P<0.05$,

$^{\#\#}P<0.01$, $^{\#\#\#}P<0.001$, $^{\#\#\#\#}P<0.0001$ and the impact of adjuvants as $*P<0.05$, $**P<0.01$, $***P<0.001$, $****P<0.0001$ (Ward et al, 2020: *Open Access: This article is licensed under a Creative Commons Attribution 4.0 International License, which permits use, sharing, adaptation, distribution and reproduction in any medium or format, as long as you give appropriate credit to the original author(s) and the source, provide a link to the Creative Commons license, and indicate if changes were made. The images or other third party material in this article are included in the article's Creative Commons license, unless indicated otherwise in a credit line to the material. If material is not included in the article's Creative Commons license and your intended use is not permitted by statutory regulation or exceeds the permitted use, you will need to obtain permission directly from the copyright holder. To view a copy of this license, visit http://creativecommons.org/licenses/by/4.0/*):

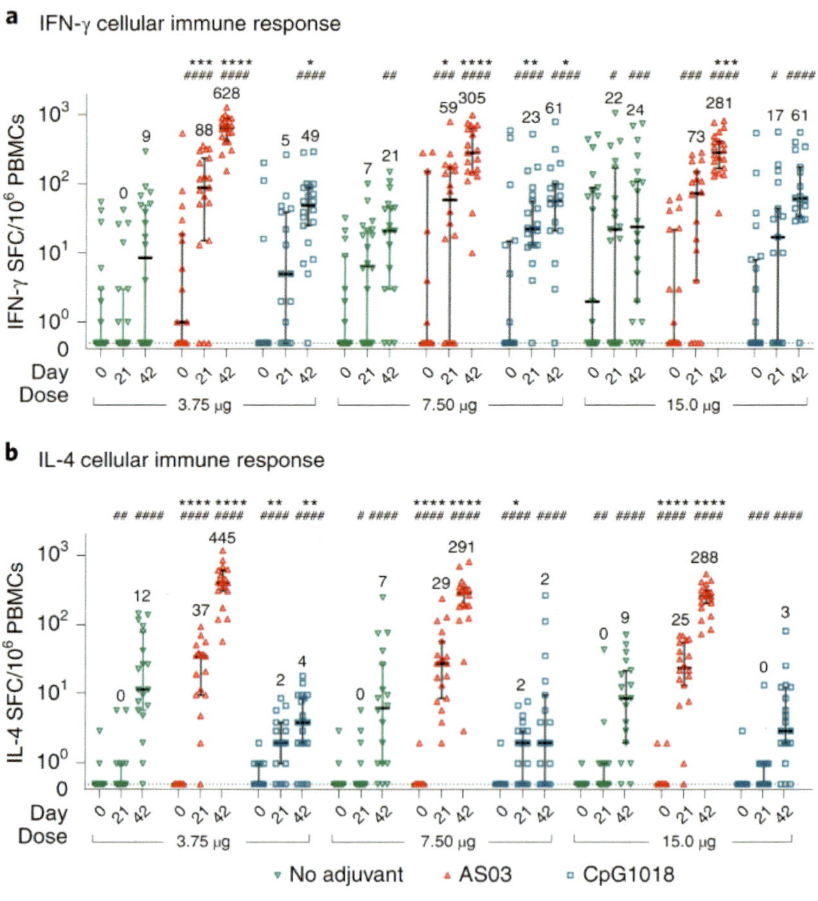

As far as safety is concerned, pain at the site of injection was the most predominant of adverse events (AEs) followed by headache and fatigue. There were no serious adverse events (SAEs) or adverse events of special interest (AESIs) reported by the authors. Figure 14 below depicts the solicited AEs: local (**a**) and systemic (**b**) with no potentially life-threatening event (Ward et al, 2020: *Open Access: This article is licensed under a Creative Commons Attribution 4.0 International License, which permits use, sharing, adaptation, distribution and reproduction in any medium or format, as long as you give appropriate credit to the original author(s) and the source, provide a link to the Creative Commons license, and indicate if changes were made. The images or other third party material in this article are included in the article's Creative Commons license, unless indicated otherwise in a credit line to the material. If material is not included in the article's Creative Commons license and your intended use is not permitted by statutory regulation or exceeds the permitted use, you will need to obtain permission directly from the copyright holder. To view a copy of this license, visit http://creativecommons.org/licenses/by/4.0/):*

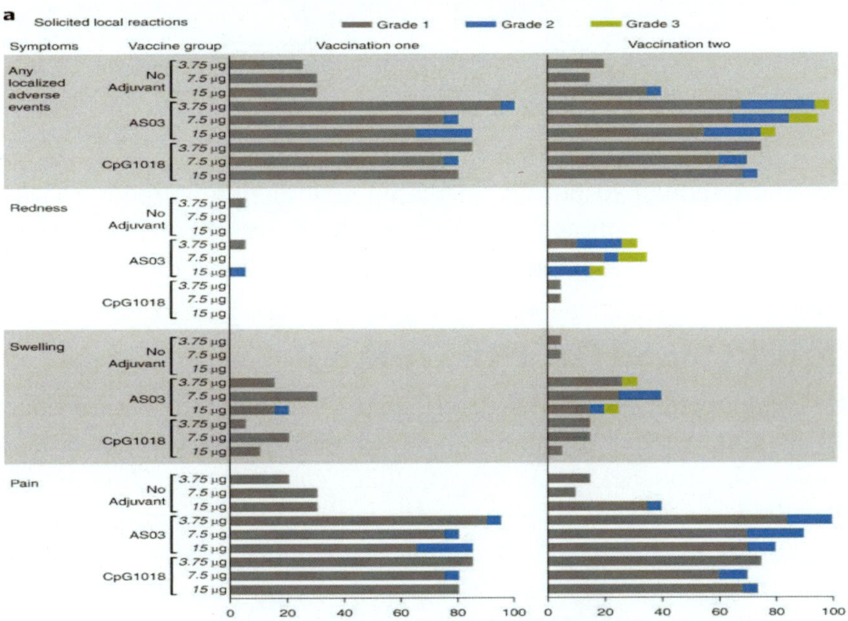

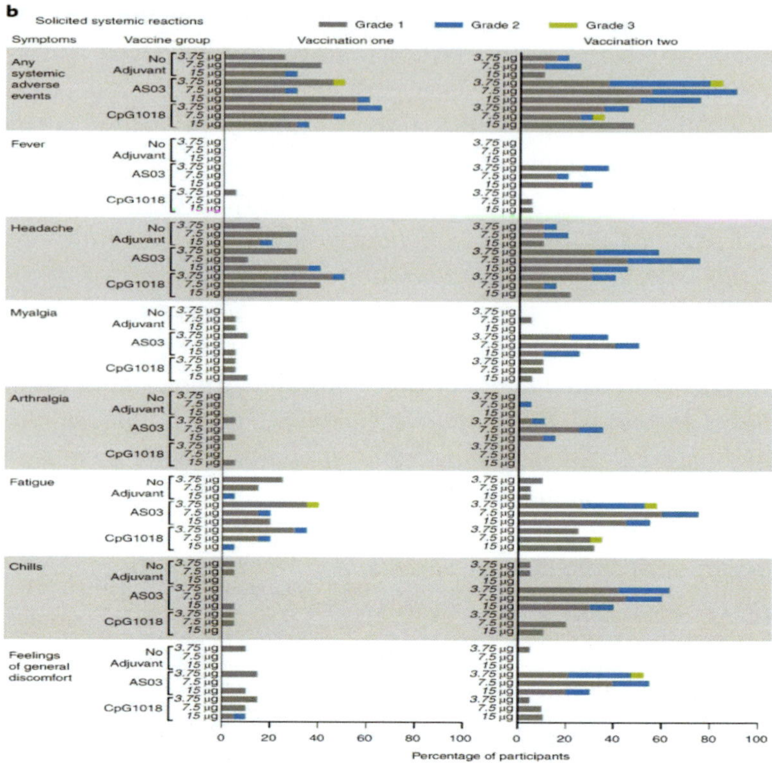

In a nutshell, the induction of spike protein-specific interferon-γ and interleukin-4 cellular responses, increased neutralizing antibody (NAb) in the presence of adjuvants and the mild to moderate impact on participants by CoVLP (NCT04450004) pave the way for detailed elucidation.

1.4 MICROBES FOR TREATMENT

A slew of applications have been described for "live biotherapeutic products" (LBP) as summarized in table 7 below based on the NIH (Cordaillat-Simmons, M., Rouanet, A. & Pot, B, 2020: *Open Access: This article is licensed under a Creative Commons Attribution 4.0 International License, which permits use, sharing, adaptation, distribution and reproduction in any medium or format, as long as you give appropriate credit to the original author(s) and the source, provide a link to the Creative Commons license, and indicate if changes were made. The images or other third party material in this article are included in the article's Creative Commons license, unless indicated otherwise in a credit line to the material. If material is not included in the article's Creative Commons license and your intended use is not permitted by statutory regulation or exceeds the permitted use, you will need to obtain permission directly from the copyright holder. To view a copy of this license, visit http://creativecommons.org/licenses/by/4.0/*):

Gastrointestinal conditions	Antibiotic-associated diarrhea *Clostridioides difficile* infection Constipation Diarrhea caused by cancer treatment Diverticular disease Inflammatory bowel disease Irritable bowel syndrome Traveler's diarrhea
Conditions in infants	Infant colic Necrotizing enterocolitis Sepsis in infants
Dental disorders	Dental caries (tooth decay) Periodontal diseases (gum disease)
Conditions related to allergy	Allergic Rhinitis (hay fever) Asthma Atopic dermatitis Prevention of allergies
Gut–brain axis related conditions	Anxiety and stress Cognition and cognitive reactivity Depression and mood Autism spectrum disorder Schizophrenia Parkinson disease Alzheimer disease
Other conditions	Acne Hepatic encephalopathy Upper respiratory infections Urinary tract infections Genital tract disorders

Interventions based on the microbiota have impacted several biomarkers associated with maladies as depicted below in table 8 (Cordaillat-Simmons, M., Rouanet, A. & Pot, B, 2020: *Open Access: This article is licensed under a Creative Commons Attribution 4.0 International License, which permits use, sharing, adaptation, distribution and reproduction in any medium or format, as long as you give appropriate credit to the original author(s) and the source, provide a link to the Creative Commons license, and indicate if changes were made. The images or other third party material in this article are included in the article's Creative Commons license, unless indicated otherwise in a credit line to the material. If material is not included in the*

Application	**Markers**
Inflammation, oxidative stress and pregnancy outcomes in gestational diabetes	Fasting plasma glucose (FPG) Serum high-sensitivity C-reactive protein (hs-CRP) Plasma malondialdehyde (MDA) concentrations MDA/TAC ratio Significant increase in total antioxidant capacity (TAC) levels
Inflammatory disorders, including allergies, diabetes, obesity, heart diseases and cancer	Global cytokines MicroRNA gene expression analyses
Aflatoxin exposure	Aflatoxin B(1)-N(7)-guanine
Oxidative stress in diabetic patients	Malondialdehyde [MDA] Glutathione [GSH] Nitric oxide [NO] Total antioxidant capacity [TAC]
MS patients	Malondialdehyde 8-Hydroxy-2′-deoxyguanosine IL-6 C-reactive protein IL-10, nitric oxide
Parkinson disease	Alpha-synuclein accumulation Dopaminergic loss in the substantia nigra pars compacta Synucleinopathy Lewy bodies

In addition to the frameworks discussed below, an LBP can achieve success in regulation and the market given that the quality, efficacy and safety all receive ticks. Live organisms as active substances are often termed "probiotics": for the public, this is a food supplement and not ubiquitously considered as "a drug". According to an FAO and WHO expert group "probiotics" are "live microorganisms which when administered in adequate amounts confer a health benefit on the host" (*Probiotics in Food: Health and Nutritional Properties and Guidelines for Evaluation* (Food and Agriculture Organization of the United Nations, Rome, 2006). With reference to regulation, the target population and the use of such a biological product can be both a drug or food product/supplement.

Figure 15 below illustrates these perspectives Cordaillat-Simmons, M., Rouanet, A. & Pot, B, 2020: *Open Access: This article is licensed under a Creative Commons Attribution 4.0 International License, which permits use, sharing, adaptation, distribution and reproduction in any medium or format, as long as you give appropriate credit to the original author(s) and the source, provide a link to the Creative Commons license, and indicate if changes were made. The images or other third party material in this article are included in the article's Creative Commons license, unless indicated otherwise in a credit line to the material. If material is not included in the article's Creative Commons license and your intended use is not permitted by statutory regulation or exceeds the permitted use, you will need to obtain permission directly from the copyright holder. To view a copy of this license, visit http://creativecommons.org/licenses/by/4.0/*):

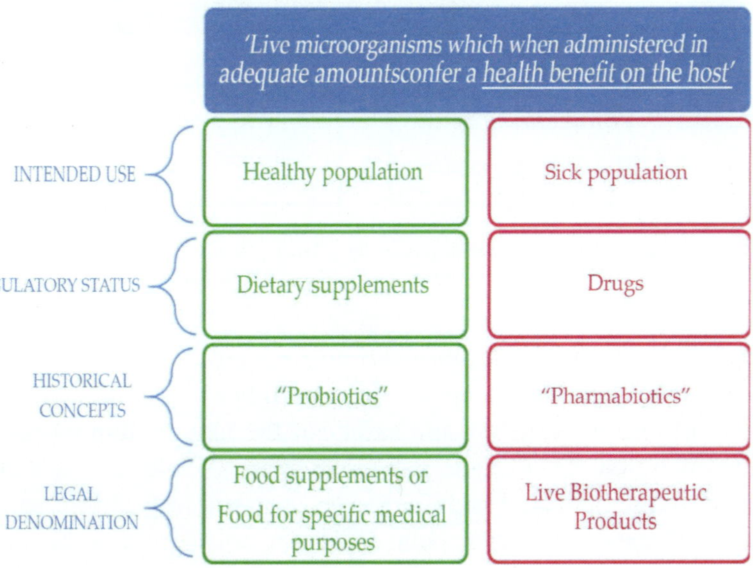

1.5 FRAMEWORKS

Crucial recommendations from the International Council for Harmonization of Technical Requirements for Pharmaceuticals for Human Use (ICH) present a platform where the requirements of regulatory authorities are put forth: European Medicines Agency (EMA) in the EU and the FDA in the US.

Figure 16 below looks at the ICH frameworks (Cordaillat-Simmons, M., Rouanet, A. & Pot, B, 2020: *Open Access: This article is licensed under a Creative Commons Attribution 4.0 International License, which permits use, sharing, adaptation, distribution and reproduction in any medium or format, as long as you give appropriate credit to the original author(s) and the source, provide a link to the Creative Commons license, and indicate if changes were made. The images or other third party material in this article are included in the article's Creative Commons license, unless indicated otherwise in a credit line to the material. If material is not included in the article's Creative Commons license and your intended use is not permitted by statutory regulation or exceeds the permitted use, you will need to obtain permission directly from the copyright holder. To view a copy of this license, visit http://creativecommons.org/licenses/by/4.0/*):

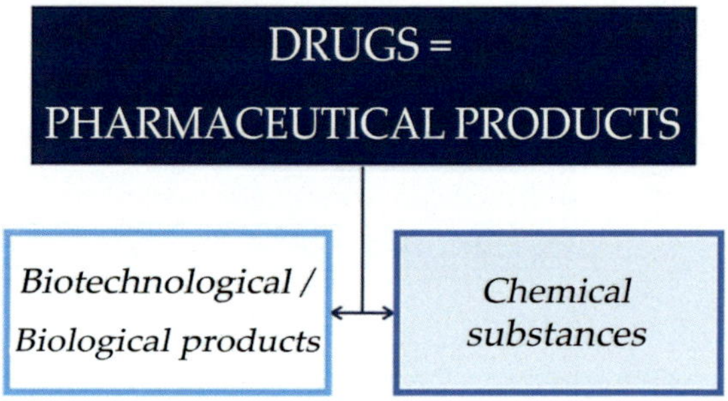

For the EU, medicinal biological products are warranted to fit in certain requirements. For instance, "The assessment of the benefits and risks in the context of a new drug application is a central element of the scientific evaluation of a marketing authorisation application (…). The assessment must reach, as objectively as possible, a sufficient level of confidence that a set level of quality, efficacy and safety of the new medicinal product has been demonstrated. This requires evaluation of all relevant data as well as the use of judgement and arguments. (…) marketing authorisation shall be refused if the benefit–risk balance is not considered to be favourable or if therapeutic efficacy is insufficiently substantiated" (*Reflection Paper on Benefit-Risk Assessment Methods in the Context of the Evaluation of Marketing Authorisation Applications of Medicinal Products for Human Use* (EMA, 2008).

For the FDA, "Simply put, for a drug to be approved for marketing, FDA must determine that the drug is effective and that its expected benefits outweigh its potential risks to patients. This assessment is informed by an extensive body of evidence about the drug's safety and efficacy, submitted

CHAPTER 2

Monoclonal Antibodies: What's New

"We went down into the basement of the institute, which has no windows," he said. "I looked at the first two plates. I saw these halos. That was fantastic. I shouted. I kissed my wife, I was all happy. It was the best result I could think of."

- Georges Kohler: co-inventor of monoclonal antibodies on the first batch of fused cells (halos indicate antibody-secreting cells) and the Nobel Prize winner with his partner César Milstein in Medicine in 1984

Contents

2.1 BACKGROUND

Plasma cells (from B-lymphocytes) produce the bodyguards called immunoglobulins (antibodies) to interact with antigens and directly inhibit the latter or recruit other immune cells to target the antigen. In our system, we have polyclonal antibodies from many plasma cells targeting a slew of antigens. In order to boost the specificity of targeting, a 'monoclonal antibody' or mAb derived from a plasma cell specifically targeting an antigen enters the stage. The integral structure of any antibody entails a Y-shaped structure of two heavy polypeptide chains and two light chains with constant (C) or variable (V) sequences. From the light and heavy chains (V_L and C_L regions; V_H and C_H1 regions) arise the antigen-binding (F_{ab}) regions of an antibody. The paratope (the site binding an antigen epitope) is on the V_H and V_L domains. The immune system is activated by the fragment crystallizable (F_c) region of the antibody that binds to a slew of immune cells. The heavy chain constant sequences are the basis for the categorization of antibodies as immunoglobin M (IgM), IgD, IgG, IgE and IgA.

Figure 1 below is illustrative of a monoclonal antibody (Baah, S., Laws, M., & Rahman, K. M: 2021 © 2021 by the authors: Licensee MDPI, Basel, Switzerland. *This article is an open access article distributed under the terms and conditions of the Creative Commons Attribution (CC BY) license: https://creativecommons.org/licenses/by/4.0/*):

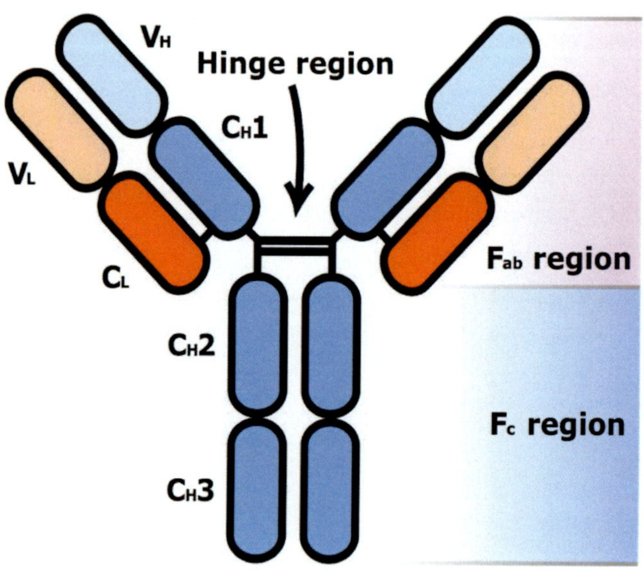

Initial work on hyperimmunised animal sera led to the employment of antibodies from mice. Such murine antibodies were associated with lowered efficacy and high immunogenicity that led to the synthesis of antibodies with human constant regions and mouse variable regions or chimeric antibodies. One can bring into mind the landmark hybridoma development of a myeloma and plasma B cell. The presence of murine sequences in these antibodies also was linked to immunogenicity that later led to murine sequences only on complementarity-determining region (CDR) loops called humanized antibodies. Subsequent advancements inclusive of cloning of the antibody sequences and phage display technology have facilitated of fully human antibodies.

Table 1 below depicts the murine, chimeric, humanized and human mAbs (mouse regions in orange and human sequences in blue) (Baah, S., Laws, M., & Rahman, K. M: 2021 © 2021 by the authors: Licensee MDPI, Basel, Switzerland. *This article is an open access article distributed under the terms and conditions of the Creative Commons Attribution (CC BY) license: https://creativecommons.org/licenses/by/4.0/*):

Type	**Suffix**	**Origin**	**Therapeutic Potential**	**Example**	**Structure**
Murine	-onab	100% derived from mouse genes (both light and heavy chains)	Problems with immunogenicity, short half-lives and limited tumor site penetration	Muromonab	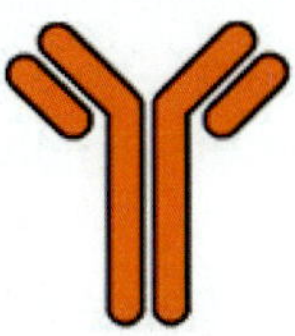
Chimeric	-ximab	35% mouse, 65% human (murine variable regions, human constant region on each chain)	Reduced immunogenicity, improved half-lives versus murine mAbs	Rituximab	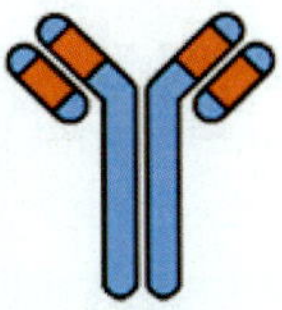
Humanised	-zumab	95% human (part of the variable domain in each chain is either synthetic or animal-derived)	Reduced immunogenicity versus chimeric mAbs	Alemtuzumab	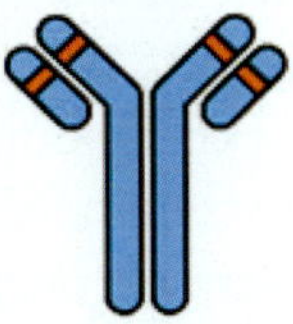

Human	-mumab	100% human (both chain types are of human origin)	Broadly reduced immunogenicity versus humanised mAbs	Adalimumab	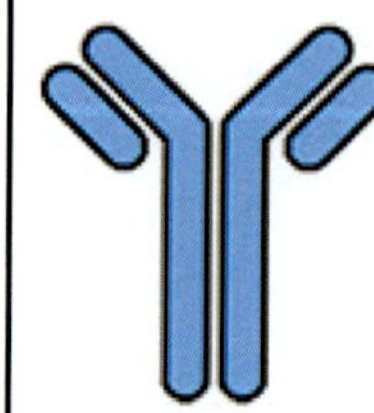

2.2 HUMANIZATION

1986 was the first US FDA approval of the murine muromonab-CD3 (Orthoclone OKT3) to address acute transplant rejection by targeting CD3. 1194 saw the approval of abciximab created by amalgamating the murine variable domain with the human constant region domain to create the first chimeric antibody for addressing platelet aggregation in cardiovascular diseases. Another chimeric antibody targeting CD20 for non-Hodgkin's lymphoma was approved in 1997 (rituximab-the first oncological target). Humanized antibodies were the outcome of grafting non-human antibody CDR sequences into a human sequence bringing in the first approval of daclizumab targeting IL-2 to address transplant rejections. Sir Gregory P. Winter is credited with fully human mAbs with phage display entailing the creation of a phage library (the approach was developed by George P. Smith) of filamentous bacteriophages with a slew of exogenous sequences. 2002 saw the first approval of a fully humanized mAb, adalimumab (Humira) for treatment targeting tumor necrosis factor α (TNFα) in rheumatoid arthritis.

Figure 2 below is an illustration of the humanization of antibodies **a** a murine monoclonal antibody. **b** a chimeric monoclonal antibody: variable regions from mice and the rest of the chains are of human origin. **c** Humanized monoclonal antibody with mice hypervariable segments. **d** Human; green shows mice sequences and orange is for human: C_H: domains of the constant region of the heavy chain; C_L: constant domain of the light chain; Fab and Fc: fragments resulting from proteolysis; V_H: variable domain of the heavy chain; V_L: variable domain of the light chain (Lu et al, 2020: *Open Access: This article is distributed under the terms of the Creative Commons Attribution 4.0 International License (http://creativecommons.org/licenses/by/4.0/), which permits unrestricted use, distribution, and reproduction in any medium, provided you give appropriate credit to the original author(s) and the source, provide a link to the Creative Commons license, and indicate* *if changes were made. The Creative Commons Public Domain Dedication waiver (http://creativecommons.org/publicdomain/zero/1.0/) applies to the data made available in this article, unless otherwise stated*):

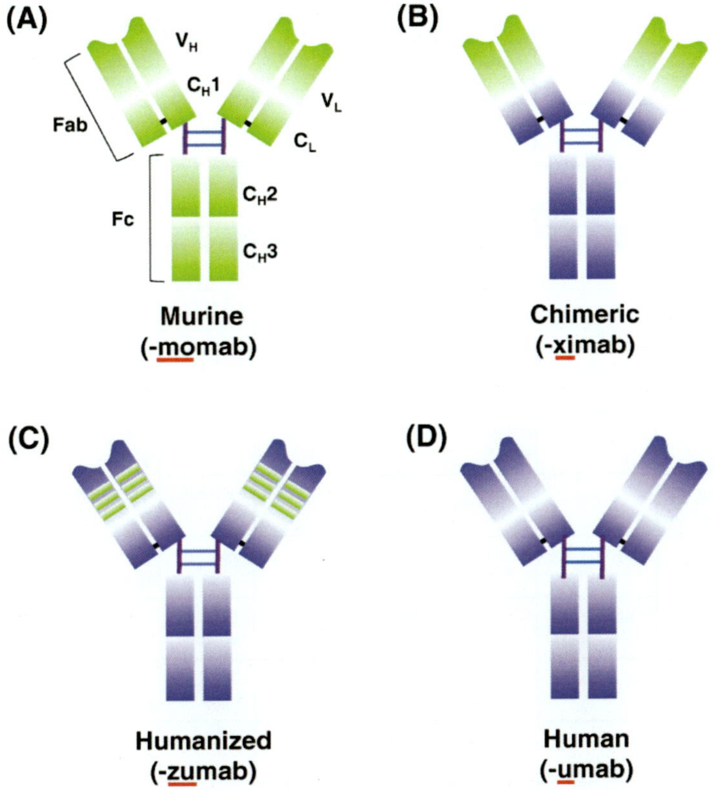

Table 2 below represents the FDA-approved human mAbs (Lu et al, 2020: *Open Access: This article is distributed under the terms of the Creative Commons Attribution 4.0 International License (http://creativecommons.org/licenses/by/4.0/), which permits unrestricted use, distribution, and reproduction in any medium, provided you give appropriate credit to the original author(s) and the source, provide a link to the Creative Commons license, and indicate if changes were made. The Creative Commons Public Domain Dedication waiver (http://creativecommons.org/publicdomain/zero/1.0/) applies to the data made available in this article, unless otherwise stated*):

No.	Antibody	Brandname	Company	Approval#	Target	References	Technology
1	Adalimumab	Humira	Abbott_Laboratories	2002	TNFα	den Broeder et al. [178]	Phage display
2	Panitumumab	Vectibix	Amgen	2006	EGFR	Tyagi et al. [179]	XenoMouse
3	Ustekinumab	Stelara	Johnson & Johnson	2009	IL-12	Bartlett et al. [17]	HuMabMouse
4	Canakinumab	Ilaris	Novartis	2009	IL-1β	Church et al. [18]	HuMabMouse
5	Golimumab	Simponi	Johnson & Johnson/Merck	2009	TNFα	Zhou et al. [19]	HuMabMouse
6	Ofatumumab	Arzerra	GlaxoSmithKline (Genmab)	2009	CD20	Coiffier et al. [20]	HuMabMouse
7	Denosumab	Prolia, Xgeva	Amgen	2010	RANKL	Reddy et al. [21]	XenoMouse
8	Belimumab	Benlysta	GlaxoSmithKline	2011	BCAF	Ding et al. [180]	Phage display
9	Ipilimumab	Yervoy	Bristol-Myers Squibb	2011	CTLA-4	Morse et al. [22]	HuMabMouse
10	Ramucirumab	Cyramza	Eli Lilly (ImClone)	2014	VEGFR2	Krupitskaya et al. [181]	Phage display
11	Nivolumab	Opdivo	Bristol-Myers Squibb	2014	PD-1	Wolchok et al. [23]	HuMabMouse
12	Alirocumab	Praluent	Sanofi and Regeneron	2015	PCSK9	Roth et al. [24]	Veloclmmune Mouse
13	Daratumumab	Darzalex	Johnson & Johnson (Genmab)	2015	CD38	de Weers et al. [25]	HuMabMouse
14	Necitumumab	Portrazza	Eli Lilly (ImClone)	2015	EGFR	Kuenen et al. [182]	Phage display
15	Evolocumab	Repatha	Amgen	2015	PCSK9	Hirayama et al. [26]	XenoMouse
16	Secukinumab	Cosentyx	Novartis	2015	IL-17α	Chioato et al. [27]	XenoMouse
17	Olaratumab	Lartruvo	Eli Lilly	2016	PDGFRα	Chiorean et al. [28]	HuMabMouse
18	Atezolizumab	Tecentriq	Roche	2016	PD-L1	McDermott et al. [183]	Phage display
19	Avelumab	Bavencio	Pfizer	2017	PD-L1	Boyerinas et al. [184]	Phage display
20	Brodalumab	Siliq	Valeant Pharmaceuticals	2017	IL-17R	Papp et al. [29]	XenoMouse
21	Dupilumab	Dupixent	Sanofi and Regeneron	2017	IL-4R	Wenzel et al. [30]	Veloclmmune Mouse
22	Durvalumab	Imfinzi	Medimmune/AstraZeneca	2017	PD-L1	Antonia et al. [31]	XenoMouse
23	Guselkumab	Tremfya	Jassen Biotech	2017	IL-23	Sofen et al. [185]	Phage display

24	Sarilumab	Kevzara	Sanofi and Regeneron	2017	IL-6R	Huizinga et al. [32]	Veloclmmune Mouse
25	Erenumab	Aimovig	Novartis and Amgen	2018	CGRPR	Tepper et al. [33]	XenoMouse
26	Cemiplimab	Libtayo	Regeneron	2018	PD-1	Migden et al. [34]	Veloclmmune Mouse
27	Emapalumab	Gamifant	NovImmmune	2018	IFNγ	Al-Salama ZT [186]	Phage display
28	Moxetumomab pasudodox	Lumoxiti	MedImmune/AstraZeneca	2018	CD22	Kreitman et al. [187]	Phage display

2.3 TRENDS

A slew of conditions employ mAbs for their treatment. Starting off by Köhler and Milstein in 1975, "magic bullets" were launched from 1981 to 1986 as depicted in figure 3 below: The height of the line with values are the estimated market value (billions of US dollars). Red antibodies are the top 10 best-selling antibody drugs in 2018: *Ab, antibody; ALCL, systematic anaplastic large-cell lymphoma; aTTP, acquired thrombotic thrombocytopenic purpura; BC, breast cancer; CD, cluster of differentiation; CGRP, calcitonin gene-related peptide; CGRPR, calcitonin gene-related peptide receptor; CRC, colorectal cancer; CTLA-4, cytotoxic T-lymphocyte-associated protein 4; EGFR, epidermal growth factor receptor; FGF, fibroblast growth factor; GC, gastric cancer; GD2, disialoganglioside G_{D2}; HER2, human epidermal growth factor receptor 2; IgE, immunoglobulin E; IL, interleukin; IL-17R, interleukin-17 receptor; mAb, monoclonal antibody; MCC, merkel-cell carcinoma; NSCLC, non-small cell lung cancer; PD-1, programmed cell death protein 1; PD-L1, programmed death-ligand 1; TNFα, tumor necrosis factor α; RA, rheumatoid arthritis; RANKL, receptor activator of nuclear factor kappa-B ligand; VEGF-A, vascular endothelial growth factor A; VEGFR2, vascular endothelial growth factor receptor 2; vWF, von Willebrand factor; XLH, X-linked hypophosphatemia* (Lu et al, 2020: *Open Access: This article is distributed under the terms of the Creative Commons Attribution 4.0 International License (http://creativecommons.org/licenses/by/4.0/), which permits unrestricted use, distribution, and reproduction in any medium, provided you give appropriate credit to the original author(s) and the source, provide a link to the Creative Commons license, and indicate if changes were made. The Creative Commons Public Domain Dedication waiver (http://creativecommons.org/publicdomain/zero/1.0/) applies to the data made available in this article, unless otherwise stated*):

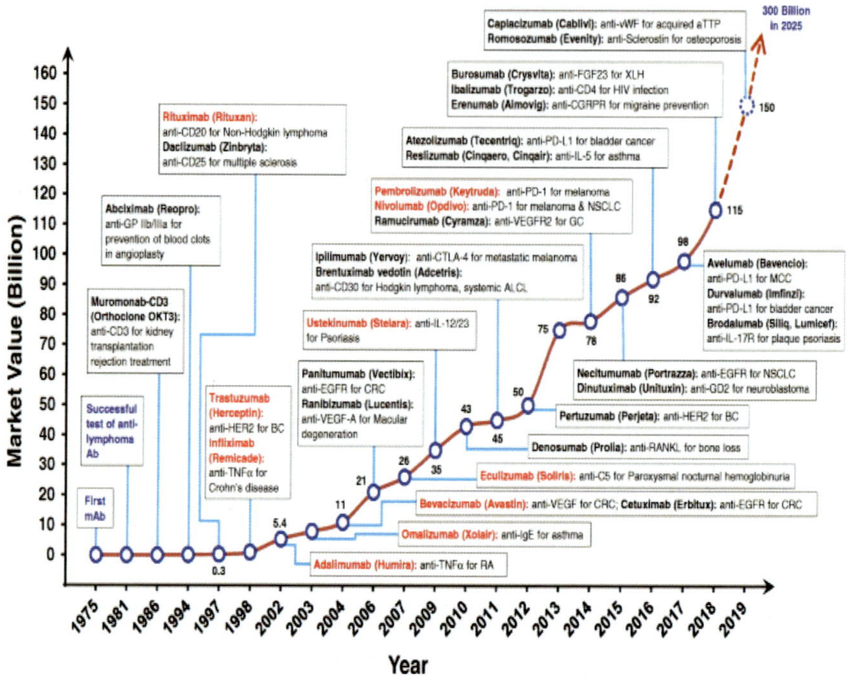

Described as a healthy pipeline, the mAb market was valued at $115.2 billion in 2018. 2018 saw six mAb products touching $6 billion sales while a slew of mAb-based products reached US$3 billion in the same year. The top ten selling mAb products of 2018 are listed in table 3 below (Lu et al, 2020: *Open Access: This article is distributed under the terms of the Creative Commons Attribution 4.0 International License (http://creativecommons.org/licenses/by/4.0/), which permits unrestricted use, distribution, and reproduction in any medium, provided you give appropriate credit to the original author(s) and the source, provide a link to the Creative Commons license, and indicate if changes were made. The Creative Commons Public Domain Dedication waiver (http://creativecommons.org/publicdomain/zero/1.0/) applies to the data made available in this article, unless otherwise stated*):

No.	Drug	Indication (1st US FDA Approval Year)	Company	2018 Revenue (USD)
1	Adalimumab (Humira)	Rheumatoid arthritis (2002) Psoriatic arthritis (2005) Ankylosing spondylitis (2006) Juvenile Idiopathic Arthitis (2008) Psoriasis (2008) Crohn's disease (2010) Ulcerative colitis (2012) Hidradenitis suppurativa (2015) Uveitis (2018)	AbbVie	$19.9 bn
2	Nivolumab (Opdivo)	Melanoma (2015) Non-small cell lung cancer (2015) Renal cell carcinoma (2015) Head and neck squamous cell (2016)	Bristol-Myers Squibb	$7.6 bn
3	Pembrolizumab (Keytruda)	Melanoma (2014) Head and neck cancer (2016) Non-small cell lung caccer (2015) Lymphoma (2018) Cervical cancer (2018) Microsatellite instability-high cancer (2018)	Merck & Co	$7.2 bn
4	Trastuzumab (Herceptin)	Breast cancer (1998) Gastric cancer (2010)	Roche	$7.0 bn
5	Bevacizumab (Avastin)	Colorectal cancer (2004) Non-small cell lung caccer (2006) Breast ERB2 negative cancer (2008) Renal cell carcinoma (2009) Glioblastoma (2011)	Roche	$6.8 bn
6	Rituximab, (Rituxan)	Non-Hodgkin's lymphoma (1997) Chronic lymphocytic leukemia (2010) Rheumatoid arthritis (2006) Pemphigus vulgaris (2018)	Roche	$6.8 bn
7	Infliximab (Remicade)	Crohn's Disease (1998) Rheumatoid arthritis (1999) Ankylosing spondylitis (2004) Ulcerative colitis (2005) Psoriatic arthritis (2005) Psoriasis (2006)	Johnson & Johnson	$5.9 bn
8	Ustekinumab (Stelara)	Psoriasis (2009) Psoriatic arthritis (2013) Crohn's Disease (2016)	Johnson & Johnson	$5.2 bn

9	Eculizumab (Soliris)	Paroxysmal nocturnal hemoglobinuria (2007) Atypical hemolytic uremic syndrome (2011) Generalized myasthenia gravis (2017) Neuromyelitis optica spectrum disorder (2019)	Alexion	$3.6 bn
10	Omalizumab (Xolair)	Asthma (2003) Chronic idiopathic urticaria (2014)	Roche	$3.0 bn

The impact of mAbs in *The Scientific World Journal* was discussed by José F. Santos-Neto et al (2021) employing the Derwent Innovation Index (DWPI) and Patent Inspiration databases. Few interesting patents between 2019 and 2020 are represented in table 4 below (José F. Santos-Neto et al, 2021: Copyright © 2021 José F. Santos-Neto et al. *This is an open access article distributed under the Creative Commons Attribution License, which permits unrestricted use, distribution, and reproduction in any medium, provided the original work is properly cited*):

Publication number	**Title**	**Publishing date**	**Description**
CN106103466 B	New blood-brain barrier shuttle comprising a brain effector entity and a brain targeting peptide, useful for treating Alzheimer's disease.	March 31, 2020	The invention provides a blood-brain barrier shuttle comprising a brain effector entity and a target brain peptide [85].
IN334250 B	New drug conjugate comprising drug moiety covalently attached to rest of drug conjugate is human epidermal growth factor receptor 2 inhibitor, useful for treating cancer, preferably lung, colorectal, breast, and pancreatic cancers.	March 13, 2020	New drug conjugates, drug-binding compounds, pharmaceutical compositions containing said drug conjugates and their use as antitumor agents. The researchers took advantage of the cell surface receptors and antigens selectively expressed by target cells, such as cancer cells, to develop pharmaceutical entities based on antibodies that bind, in the example of tumors, specific or tumor-associated antigens [86].

ES2742351T3	New isolated antibody or its immunoreactive fragment comprises a variable domain that specifically binds an extracellular domain of B7-H3, useful for diagnosing cancer, and for preparing a medicament for treating cancer in a patient.	February 14, 2020	Antibodies and their fragments that are immunoreactive against the mammalian, and more particularly human, B7-H3 receptor and uses thereof, particularly in the treatment of cancer and inflammation. The invention therefore relates particularly to humanized B7-H3 reactive antibodies and their immunoreactive fragments that can mediate and more preferably enhance activation of the immune system against cancer cells that are associated with a variety of human cancers [87].
JP06642946 B2	New ligand drug conjugate compound comprising beta-glucuronide based linker for treating cancer, autoimmune disease, and infectious disease.	February 12, 2020	A ligand drug conjugate comprising a ligand, such as an antibody, provided for allowing the conjugate to target a target cell or target tissue. The conjugate further comprises a β-glucuronide-based linker including a site that can be cleaved by an enzyme having β-glucuronidase activity. The linker is bound to the ligand and the drug. According to the present invention, furthermore, there is provided a method for treating cancers, immune diseases, infectious diseases, and other diseases or disorders using the ligand drug conjugate comprising the β-glucuronide-based linker [88].
AU2020200175 A1	New antibody-drug conjugate compound where a linker drug is site-specifically conjugated to an antibody through an engineered cysteine for use in treating solid tumors and hematological malignancies, e.g., breast cancer and prostate cancer.	January 30, 2020	Relates to antibody-drug conjugates (ADCs), wherein a linker drug is site-specifically conjugated to an antibody through an engineered cysteine, and their use as a medicament, notably for the treatment of human solid tumors and hematological malignancies, in particular breast cancer, gastric cancer, colorectal cancer, urothelial cancer, ovarian cancer, uterine cancer, lung cancer, mesothelioma, liver cancer, pancreatic cancer, prostate cancer, and leukemia [89].

CA2665480C	New peptide conjugate comprising a peptide which is covalently attached to a moiety, useful for preparing a composition for treating tumor.	November 11, 2019	Production of conjugates between peptides and PEG moieties through glycerol linkers. In one aspect, the present invention provides a conjugate between a PEG moiety, e.g., PEG and a peptide that has an in vivo activity similar or otherwise analogous to art-recognized therapeutic peptide. In the conjugate of the invention, the PEG moiety is covalently attached to the peptide via an intact glycosyl linking group. Exemplary intact glycosyl linking groups include sialic acid moieties that are derivatized with PEG [90].
BRPI0608281 B1	The human monoclonal antibody or its fragment, binding to and neutralizing primate granulocyte-macrophage colony-stimulating factor, useful for treating inflammatory diseases and cancer.	January 28, 2020	A human monoclonal antibody or fragment of its own specifically binds GM-CSF and neutralizes it [91].
NO344249 B1	The human monoclonal antibody or its fragment, binding to and neutralizing primate granulocyte-macrophage colony-stimulating factor, useful for treating inflammatory diseases and cancer.	October 21, 2019	A human monoclonal antibody or fragment that specifically binds to and neutralizes primate GM-CSF [92].

Further analyses of the Web of Science (WoS) database revealed the following prospective mAb-based works (José F. Santos-Neto et al, 2021:):

Title	Journal	Publication year	Description
Antitumor effect of humanized antitissue factor antibody-drug conjugate in a model of peritoneal disseminated pancreatic cancer	Oncology Reports	2020	The work evaluated the action of an original humanized anti-TF monoclonal antibody conjugated with monomethyl auristatin E against the tissue factor (TF), an attractive target involved in the overexpression of a wide variety of malignant cancers, through in vivo efficacy trials [111]
Acquired resistance to HER2-targeted therapies creates vulnerability to ATP synthase inhibition	Cancer Research	2020	The work evaluates the effectiveness of the combination of ATP oligomycin A inhibitor with trastuzumab oligomycin A in order to lead to the regression of tumors resistant to this antibody, such as HER2+ breast cancer patients [112]
InsB9-23 gene transfer to hepatocyte-based combined therapy abrogates recurrence of type 1 diabetes after islet transplantation	Diabetes	2021	The work evaluates a treatment option for autoimmune diabetes by using a lentiviral vector that allows the expression of the insulin B chain 9–23 in combination with a suboptimal dose of anti-CD3 monoclonal antibody [113]
Efficacy and safety of bevacizumab in advanced lung adenocarcinoma patients with stable disease after two cycles of first-line chemotherapy: A multicenter prospective cohort study	Thoracic Cancer	2020	This work is based on a clinical trial that aimed to evaluate the effects after treatment of patients with advanced pulmonary adenocarcinoma after two cycles of platinum-based double agent chemotherapy compared to patients with the same disease and same treatment but in association with bevacizumab [114]
Computational engineering the binding affinity of adalimumab monoclonal antibody for designing potential biosimilar candidate	Journal of Molecular Graphics and Modeling	2021	The work addresses a modeling study using bioinformatics tools to find variants of the adalimumab monoclonal antibody with improved binding properties since some risk factors are associated with high dosages of this antibody in treatments of diseases such as rheumatoid arthritis and other autoimmune diseases [39]

Development of a dinutuximab delivery system using silk foams for GD2 targeted neuroblastoma cell death	Journal of Biomedical Materials Research Part A	2020	The work evaluates the action of dinutuximab as a treatment for patients with high-risk neuroblastoma, from a delivery system capable of delivering this bioactive antibody locally, using silk fibroin. This delivery system can potentially reduce the probability of cancer recurrence, as well as minimize side effects [115]
Treatment with tocilizumab for patients with COVID-19 infections: A case-series study	Journal of Clinical Pharmacology	2020	The work is based on an evaluation of the clinical impacts of the use of tocilizumab in the treatment of COVID-19 patients, based on a comparison between patients who received monoclonal antibody as treatment and patients who received other therapeutic alternatives [116]
Anti-complement C5 therapy with eculizumab in three cases of critical COVID-19	Clinical Immunology	2020	The work evaluated an alternative for the treatment of critical patients at COVID-19. The treatment is based on the use of a monoclonal antibody anticomplement, the anti-C5 antibody eculizumab, in the reduction of severe clinical manifestations such as respiratory failure and acute kidney damage [117]

Despite the wreak by COVID-19, therapeutic monoclonal antibodies approved by FDA in 2020 are depicted below in table 5 (Cai, 2021:):

Table 1: Therapeutic monoclonal antibodies approved by FDA in 2020

Drug name	Active ingredients	Company	Approval date
TEPEZZ	Teprotumumab-trb	HORIZON THERAPEUTICS	January 21, 2020
VYEPTI	Eptinezumab-jjmr	Lundbeck	February 21, 2020
SARCLISA	Isatuximab-irfc	SANOFI AVENTIS	March 2, 2020
TRODELVY	Sacituzumab govitecan-hziy	IMMUNOMEDICS	April 22, 2020
DARZALEX FASPRO	Daratumumab and hyaluronidase-fihj	JANSSEN	May 1, 2020
UPLIZNA	Inebilizumab-cdon	VIELA	June 11, 2020
PHESGO	Pertuzumab, trastuzumab, and hyaluronidase-zzxf	GENENTECH	June 29, 2020
BLENREP	Belantamab mafodotin-blmf	GLAXOSMITHKLINE	August 5, 2020
ENSPRYNG	Satralizumab-mwge	GENENTECH	August 14, 2020
INMAZEB	ATOLTIVIMAB +2	REGENERON	October 14, 2020
DANYELZA	NAXITAMAB-GQGK	Y-MABS THERAPEUTICS	November 25, 2020
MARGENZA	MARGETUXIMAB-CMKB	MACROGENICS	December 16, 2020
RIABNI	RITUXIMAB-ARRX	AMGEN	December 17, 2020

ATOLTIVIMAB +2 ATOLTIVIMAB; ODESIVIMAB; MAFTIVIMAB

Table 2: Some characteristics of the therapeutic Mab approved by FDA in 2020

Drug name	Indications and usage	Warnings and precautions	Mechanism of action
TEPEZZ	Thyroid eye disease	IBD worsening; hyperglycemia	IGF-1R block
VYEPTI	Migraine	Hypersensitivity	CGRP block
SARCLISA	Multiple myeloma	Neutropenia; SPM	Anti-CD38, ADCC, CDC,
TRODELVY	mTNBC	Hypersensitivity, neutropenia	Anti-Trop-2; SN-38; ADC
DARZALEX FASPRO	Multiple myeloma	Hypersensitivity, Neutropenia	Anti-CD38, ADCC, CDC,
UPLIZNA	NMOSD (AQP4+)	Immune compromise	Anti-CD19
PHESGO	Breast cancer (HER2+)	Neutropenia	Anti-HER2; increases permeability
BLENREP	Multiple Myeloma	Thrombocytopenia	Anti-BCMA; microtubule inhibitor; ADC
ENSPRYNG	NMOSD (AQP4+)	Elevated liver enzymes	Anti-IL6 receptor
INMAZEB	Zaire ebolavirus infection.	Hypersensitivity	Zaire ebolavirus glycoprotein
DANYELZA	Neuroblastoma	Neurotoxicity; Hypertension	Anti-glycolipid GD2, CDC, ADCC
MARGENZA	Breast cancer	Left ventricular dysfunction	Anti-HER2, ADCC
RIABNI	NHL, CLL, GPA, MPA	IRR; TLS; PML	Anti-CD20, ADCC, CDC

IBD: Inflammatory bowel disease, IGF-1R: Insulin-like growth factor 1 (IGF-1) receptor, CGRP: Calcitonin gene-related peptide, SPM: Second primary malignancies, ADCC: Antibody-dependent cell-mediated cytotoxicity, CDC: Complement-dependent cytotoxicity, ADCP: Antibody-dependent cellular phagocytosis, mTNBC: Metastatic triple-negative breast cancer, SN-38: Topoisomerase I inhibitor, ADC: Antibody drug conjugate, NMOSD: Neuromyelitis optica spectrum disorder, AQP4+: Anti-aquaporin-4 antibody positive, NHL: Non-Hodgkin's lymphoma, CLL: Chronic lymphocytic leukemia, GPA: Granulomatosis with polyangiitis (Wegener's granulomatosis), MPA: Microscopic polyangiitis, IRR: Infusion-related reaction, TLS: Tumor lysis syndrome, PML: Progressive multifocal leukoencephalopathy

Table 6 below is an exhaustive list of US FDA-approved mAbs in the market (Lu et al, 2020: *Open Access: This article is distributed under the terms of the Creative Commons Attribution 4.0 International License (http://creativecommons.org/licenses/by/4.0/), which permits unrestricted use, distribution, and reproduction in any medium, provided you give appropriate credit to the original author(s) and the source, provide a link to the Creative Commons license, and indicate if changes were made. The Creative Commons Public Domain Dedication waiver (http://creativecommons.org/publicdomain/zero/1.0/) applies to the data made available in this article, unless otherwise stated*):

Key: *Marketing end date on July 30th, 2011; #Year of the first US FDA approval; &Indication of the first US FDA approval; $Rabbit hybridoma technology

Ab	Brand name	Company	Target	Format	Technology	Indication&	US# Approval
Muromon-ab-CD3	Ortho-clone OKT3	Centocor Ortho Biotech Products LP.	CD3	Murine IgG2a	Hybridoma/ Janssen Bio-tech, Inc	Kidney transplant rejection	1986*
Abciximab	Reopro	Centocor Inc./ Eli Lilly/Jans-sen Biotech Inc.	GPIIb/IIIa	Chimeric IgG1 Fab	Hybridoma	Prevention of blood clots in angioplasty	1994

Rituximab	Mab-Thera, Rituxan	Biogen Inc./ Roche, F. Hoffmann-La Roche Ltd./ Genentech Inc.	CD20	Chimeric IgG1	Hybridoma	Non-Hodgkin lymphoma	1997
Palivizumab	Synagis	MedImmune/ AbbVie Inc.	RSV	Human-ized IgG1	Hybridoma	Prevention of respiratory syncytial vi-rus infection	1998
Infliximab	Remicade	Janssen Bio-tech Inc.	TNFα	Chimeric IgG1	Hybridoma	Crohn's disease	1998
Trastuzumab	Herceptin	Roche, F. Hoffmann-La Roche, Ltd./ Genentech Inc.	HER2	Human-ized IgG1	Hybridoma	Breast cancer	1998
Alemtu-zumab	Campath, Lemtrada	Berlex Inc./ Genzyme Corp./Millen-nium Pharma-ceuticals Inc.	CD52	Human-ized IgG1	Hybridoma	Chronic myeloid leukemia	2001
Adalimumab	Humira	AbbVie Inc.	TNFα	Human IgG1	Phage display	Rheumatoid arthritis	2002
Ibritumomab tiuxetan	Zevalin	Biogen Inc./ Schering AG/ Spectrum Pharmaceuti-cals Inc.	CD20	Murine IgG1	Hybridoma	Non-Hodgkin lymphoma	2002
Omalizumab	Xolair	Roche, F. Hoffmann-La Roche, Ltd./ Genentech Inc./Novartis Pharmaceu-ticals Corp./ Tanox Inc.	IgE	Human-ized IgG1	Hybridoma	Asthma	2003
Cetuximab	Erbitux	Bristol-Myers Squibb/Merck & Co. Inc./Eli Lilly/ImClone Systems Inc.	EGFR	Chimeric IgG1	Hybridoma	Colorectal cancer	2004
Bevaci-zumab	Avastin	Roche, F. Hoffmann-La Roche, Ltd./ Genentech Inc.	VEGF-A	Human-ized IgG1	Hybridoma	Colorectal cancer	2004
Natalizumab	Tysabri	Biogen Inc./ Elan Phar-maceuticals International, Ltd.	ITGA4	Human-ized IgG4	Hybridoma	Multiple sclerosis	2004
Panitu-mumab	Vectibix	Amgen	EGFR	Human IgG2	Transgenic mice	Colorectal cancer	2006

Ranibizumab	Lucentis	Roche, F. Hoffmann-La Roche Ltd./ Genentech Inc./Novartis Pharmaceuticals Corp.	VEGF-A	Humanized IgG1 Fab	Hybridoma	Macular degeneration	2006
Eculizumab	Soliris	Alexion Pharmaceuticals Inc.	C5	Humanized IgG2/4	Hybridoma	Paroxysmal nocturnal hemoglobinuria	2007
Certolizumab pegol	Cimzia	Celltech, UCB.	TNFα	Humanized Fab, pegylated	Hybridoma	Crohn's disease	2008
Ustekinumab	Stelara	Medarex/ Centocor Ortho Biotech Inc./Janssen Biotech Inc.	IL-12/23	Human IgG1	Transgenic mice	Psoriasis	2009
Canakinumab	Ilaris	Novartis Pharmaceuticals Corp.	IL-1β	Human IgG1	Transgenic mice	Muckle-Wells syndrome	2009
Golimumab	Simponi	Centocor Ortho Biotech Inc./Janssen Biotech Inc.	TNFα	Human IgG1	Transgenic mice	Rheumatoid and psoriatic arthritis, ankylosing spondylitis	2009
Ofatumumab	Arzerra	Genmab A/S /GlaxoSmithKline / Novartis.	CD20	Human IgG1	Transgenic mice	Chronic lymphocytic leukemia	2009
Tocilizumab	Ro-Actemra, Actemra	Chugai Pharmaceutical Co., Ltd./ Roche, F. Hoffmann-La Roche. Ltd./ Genentech Inc.	IL-6R	Humanized IgG1	Hybridoma	Rheumatoid arthritis	2010
Denosumab	Xgeva, Prolia	Amgen	RANKL	Human IgG2	Transgenic mice	Bone loss	2010
Belimumab	Benlysta	GlaxoSmithKline /Human Genome Sciences Inc.	BLyS	Human IgG1	Phage display	Systemic lupus erythematosus	2011
Ipilimumab	Yervoy	Bristol-Myers Squibb/Medarex	CTLA-4	Human IgG1	Transgenic mice	Metastatic melanoma	2011

Brentuximab vedotin	Adcetris	Seattle genetics Inc./Takeda Pharmaceutical Co., Ltd.	CD30	Chimeric IgG1; ADC	Hybridoma	Hodgkin lymphoma, systemic anaplastic large cell lymphoma	2011
Pertuzumab	Perjeta	Roche, F. Hoffmann-La Roche, Ltd./ Genentech Inc.	HER2	Humanized IgG1	Hybridoma	Breast Cancer	2012
Trastuzumab emtansine	Kadcyla	Roche, F. Hoffmann-La Roche Ltd./ Genentech Inc./Immuno-Gen Inc.	HER2	Humanized IgG1; ADC	Hybridoma	Breast cancer	2012
Raxibacumab	Abthrax	GlaxoSmith-Kline /Human Genome Sciences Inc. (HGSI)	*B. anthrasis* PA	Human IgG1	Transgenic mice	Anthrax infection	2012
Obinutuzumab	Gazyva, Gazyvaro	Biogen Inc./ Roche, F. Hoffmann-La Roche, Ltd./ Genentech Inc.	CD20	Humanized IgG1 Glycoengineered	Hybridoma	Chronic lymphocytic leukemia	2013
Siltuximab	Sylvant	Centocor Inc./Janssen Biotech Inc./ Janssen-Cilag International NV	IL-6	Chimeric IgG1	Hybridoma	Castleman disease	2014
Ramucirumab	Cyramza	Eli Lilly/Im-Clone Systems Inc.	VEGFR2	Human IgG1	Phage display	Gastric cancer	2014
Vedolizumab	Entyvio	Genentech Inc./Millennium Pharmaceuticals Inc./ Takeda Pharmaceuticals U.S.A. Inc.	α4β7 integrin	Humanized IgG1	Hybridoma	Ulcerative colitis, Crohn disease	2014
Blinatumomab	Blincyto	Amgen	CD19, CD3	Murine bispecific tandem scFv	Hybridoma	Acute lymphoblastic leukemia	2014
Nivolumab	Opdivo	Bristol-Myers Squibb/Ono Pharmaceutical Co., Ltd.	PD-1	Human IgG4	Transgenic mice	Melanoma, non-small cell lung cancer	2014

Pembrolizumab	Keytruda	Merck & Co. Inc.	PD-1	Humanized IgG4	Hybridoma	Melanoma	2014
Idarucizumab	Praxbind	Boehringer Ingelheim Pharmaceuticals	Dabigatran	Humanized Fab	Hybridoma	Reversal of dabigatran-induced anticoagulation	2015
Necitumumab	Portrazza	Eli Lilly/ImClone Systems Inc.	EGFR	Human IgG1	Phage display	Non-small cell lung cancer	2015
Dinutuximab	Unituxin	United Therapeutics Corporation	GD2	Chimeric IgG1	Hybridoma	Neuroblastoma	2015
Secukinumab	Cosentyx	Novartis Pharmaceuticals Corp.	IL-17α	Human IgG1	Transgenic mice	Psoriasis	2015
Mepolizumab	Nucala	Centocor Inc./GlaxoSmithKline	IL-5	Humanized IgG1	Hybridoma	Severe eosinophilic asthma	2015
Alirocumab	Praluent	Regeneron Pharmaceuticals Inc./Sanofi.	PCSK9	Human IgG1	Transgenic mice	High cholesterol	2015
Evolocumab	Repatha	Amgen/Amgen Astellas BioPharma K.K.	PCSK9	Human IgG2	Transgenic mice	High cholesterol	2015
Daratumumab	Darzalex	Genmab A/S/Janssen Biotech Inc.	CD38	Human IgG1	Transgenic mice	Multiple myeloma	2015
Elotuzumab	Empliciti	Bristol-Myers Squibb/AbbVie Inc.	SLAMF7	Humanized IgG1	Hybridoma	Multiple myeloma	2015
Ixekizumab	Taltz	Eli Lilly	IL-17α	Humanized IgG4	Hybridoma	Psoriasis	2016
Reslizumab	Cinqaero, Cinqair	Celltech, UCB/Schering-Plough/Teva Pharmaceutical Industries, Ltd.	IL-5	Humanized IgG4	Hybridoma	Asthma	2016
Olaratumab	Lartruvo	Eli Lilly/ImClone Systems Inc.	PDGFRα	Human IgG1	Transgenic mice	Soft tissue sarcoma	2016
Bezlotoxumab	Zinplava	Merck & Co. Inc.	*Clostridium difficile* enterotoxin B	Human IgG1	Transgenic mice	Prevention of *Clostridium difficile* infection recurrence	2016

Atezolizumab	Tecentriq	Roche, F. Hoffmann-La Roche, Ltd./ Genentech Inc.	PD-L1	Humanized IgG1	Hybridoma	Bladder cancer	2016
Obiltoxaximab	Anthim	Elusys Therapeutics Inc.	*B. anthrasis* PA	Chimeric IgG1	Hybridoma	Prevention of inhalational anthrax	2016
Inotuzumab ozogamicin	Besponsa	Wyeth Pharmaceuticals/ Pfizer.	CD22	Humanized IgG4	Hybridoma	Acute lymphoblastic leukemia	2017
Brodalumab	Siliq, Lumicef	MedImmune/ Amgen/Kyowa Hakko Kirin / AstraZeneca/ Valeant Pharmaceuticals International Inc.	IL-17R	Human IgG2	Transgenic mice	Plaque psoriasis	2017
Guselkumab	Tremfya	MorphoSys/ Janssen Biotech Inc.	IL-23 p19	Human IgG1	Phage display	Plaque psoriasis	2017
Dupilumab	Dupixent	Regeneron Pharmaceuticals Inc./ Sanofi	IL-4Rα	Human IgG4	Transgenic mice	Atopic dermatitis	2017
Sarilumab	Kevzara	Regeneron Pharmaceuticals Inc./ Sanofi	IL-6R	Human IgG1	Transgenic mice	Rheumatoid arthritis	2017
Avelumab	Bavencio	Merck Serono International S.A./Pfizer	PD-L1	Human IgG1	Phage display	Merkel cell carcinoma	2017
Ocrelizumab	Ocrevus	Biogen Inc./ Roche, F. Hoffmann-La Roche, Ltd./ Genentech Inc./SIGMA-TAU Industrie Farmaceutiche Riunite S.p.A.	CD20	Humanized IgG1	Hybridoma	Multiple sclerosis	2017
Emicizumab	Hemlibra	Chugai Pharmaceutical Co., Ltd./ Roche, F. Hoffmann-La Roche, Ltd.	Factor IXa, X	Humanized IgG4, bispecific	Hybridoma	Hemophilia A	2017
Benralizumab	Fasenra	MedImmune/ Kyowa Hakko Kirin/AstraZeneca	IL-5Rα	Humanized IgG1	Hybridoma	Asthma	2017

Gemtuzumab ozogamicin	Mylotarg	Pfizer	CD33	Humanized IgG4; ADC	Hybridoma	Acute myeloid leukemia	2017
Durvalumab	Imfinzi	MedImmune/ AstraZeneca	PD-L1	Human IgG1	Transgenic mice	Bladder cancer	2017
Burosumab	Crysvita	Kyowa Hakko Kirin/Ultragenyx Pharmaceutical Inc.	FGF23	Human IgG1	Transgenic mice	X-linked hypophosphatemia	2018
Lanadelumab	Takhzyro	Dyax Corp.	Plasma kallikrein	Human IgG1	Phage display	Hereditary angioedema attacks	2018
Mogamulizumab	Poteligeo	Kyowa Hakko Kirin	CCR4	Humanized IgG1	Hybridoma	Mycosis fungoides or Sézary syndrome	2018
Erenumab	Aimovig	Novartis	CGRPR	Human IgG2	Transgenic mice	Migraine prevention	2018
Galcanezumab	Emgality	Eli Lilly	CGRP	Humanized IgG4	Hybridoma	Migraine prevention	2018
Tildrakizumab	Ilumya	Merck & Co. Inc./Sun Pharmaceutical Industries, Ltd.	IL-23 p19	Humanized IgG1	Hybridoma	Plaque psoriasis	2018
Cemiplimab	Libtayo	Regeneron Pharmaceuticals Inc.	PD-1	Human mAb	Transgenic mice	Cutaneous squamous cell carcinoma	2018
Emapalumab	Gamifant	NovImmmune	IFNγ	Human IgG1	Phage display	Primary hemophagocytic lymphohistiocytosis	2018
Fremanezumab	Ajovy	Teva Pharmaceutical Industries, Ltd.	CGRP	Humanized IgG2	Hybridoma	Migraine prevention	2018
Ibalizumab	Trogarzo	Taimed Biologics Inc./ Theratechnologies Inc.	CD4	Humanized IgG4	Hybridoma	HIV infection	2018
Moxetumomab pasudodox	Lumoxiti	MedImmune/ AstraZeneca	CD22	Murine IgG1 dsFv	Phage display	Hairy cell leukemia	2018
Ravulizumab	Ultomiris	Alexion Pharmaceuticals Inc.	C5	humanized IgG2/4	Hybridoma	Paroxysmal nocturnal hemoglobinuria	2018

Caplacizumab	Cablivi	Ablynx	von Willebrand factor	Humanized Nanobody	Hybridoma	Acquired thrombotic thrombocytopenic purpura	2019
Romosozumab	Evenity	Amgen/UCB	Sclerostin	Humanized IgG2	Hybridoma	Osteoporosis in post-menopausal women at increased risk of fracture	2019
Risankizumab	Skyrizi	Boehringer Ingelheim Pharmaceuticals/ AbbVie Inc.	IL-23 p19	Humanized IgG1	Hybridoma	Plaque psoriasis	2019
Polatuzumab vedotin	Polivy	Roche, F. Hoffmann-La Roche, Ltd.	CD79β	Humanized IgG1 ADC	Hybridoma	Diffuse large B-cell lymphoma	2019
Brolucizumab	Beovu	Novartis Pharmaceuticals Corp.	VEGF-A	Humanized scFv	Hybridoma[$]	Macular degeneration	2019
Crizanlizumab	Adakveo	Novartis Pharmaceuticals Corp.	P-selectin	Humanized IgG2	Hybridoma	Sickle cell disease	2019

2.4 MABS FOR COVID-19:

A slew of therapeutic approaches are being pursued for addressing Coronavirus disease 2019 (COVID-19) caused by severe acute respiratory syndrome coronavirus 2 (SARS-CoV-2) that is inclusive of antivirals, corticosteroids, convalescent plasma and antibodies, to name a few. Passive immunotherapy is being explored: this entails convalescent plasma therapy (CPT) where a patient receives plasma from a hyperimmune patient while the other approach is the administration of antibodies (polyclonal or mAbs). The latter emerges as more specific and few have been employed off-label (non-officially approved condition) in the pandemic. Several specific COVID-19 specific/non-specific mAbs are under scrutiny.

Figure 4 below displays ***TOP***: mAb approach for COVID-19 and ***BOTTOM***: mAbs for COVID-19 (Torrente-López A, Hermosilla J, Navas N, Cuadros-Rodríguez L, Cabeza J, Salmerón-García A: © 2021 by the authors. Licensee MDPI, Basel, Switzerland. *This article is an open access article distributed under the terms and conditions of the Creative Commons Attribution (CC BY) license: https://creativecommons.org/licenses/by/4.0/*):

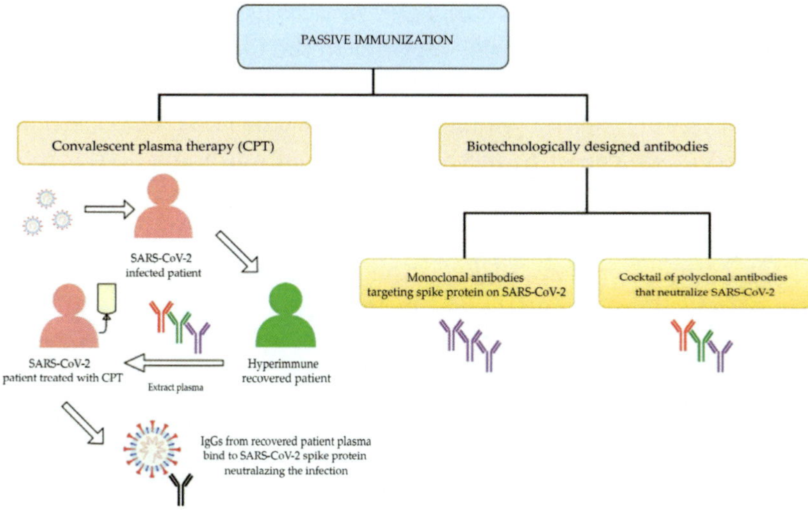

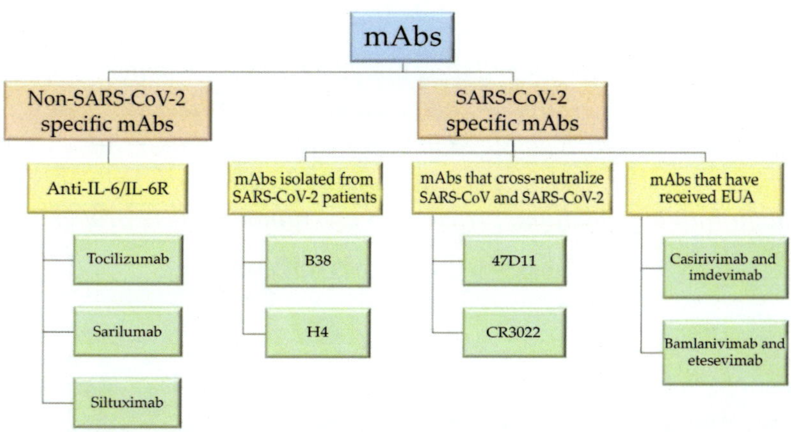

Research by Bryan E. Jones and team (2021) in Science Translational Medicine reported the high-throughput scrutiny of the antibodies targeting the surface spike protein of SARS-CoV-2 in a rhesus macaque challenge model. Anti-spike antibodies were scored employing a multiplexed bead–based assay with trimeric SARS-CoV-2 or SARS-CoV-1 spike protein a live mammalian cell–based assay with transient expression of full-length membrane-anchored SARS-CoV-2 spike protein with peripheral blood mononuclear cells (PBMCs) of a convalescent patient. Subsequent analyses unveiled a greater potency (10-fold) to neutralize the antigen in a mAb, Ab169 (named LY-CoV555) that binds to receptor-binding domain (RBD) and block angiotensin-converting enzyme 2 (ACE2).

Figure 5 below is illustrative of ***TOP***: *in vitro* assays for (**A**) neutralizing SARS-CoV-2 via a nanoluciferase reporter in the Orf7a/b locus (Luc-Virus) in infected Vero E6 cells (**B** and **C**) Plaque reduction neutralization test (PRNT) assays for INMI-1 isolate (B) and 2020/USA/WA-1 isolate (C) of SARS-CoV-2 in Vero E6: mAb, monoclonal antibody; RBD, receptor-binding domain; NTD, N-terminal domain; ***BOTTOM***: (**A**) Crystal structure of the RBD-LY-CoV555 complex superimposed with the ACE2 receptor (**B** and **C**) Zoomed-in view of key atomic interactions at the interface of the LY-CoV555 light chain (B) and heavy chain (C) with the spike RBD. (**D**) Cryo-EM structure of the LY-CoV555 with all three Fabs (in cyan). (**E**) High-resolution cryo-EM map of the LY-CoV555-spike complex. (**F**) SARS-CoV-2 RBD molecular surface with ACE2 (gray), LY-CoV555 (cyan), ACE2 and LY-CoV555 (pink): Cryo-EM, cryo–electron microscopy; RBD, receptor-binding domain (Bryan E. Jones et al, 2021: *Copyright © 2021, Copyright © 2021 The Authors, some rights reserved; exclusive licensee American Association for the Advancement of Science. No claim to original U.S. Government Works. Distributed under a Creative Commons Attribution License 4.0 (CC BY).. https://creativecommons.org/licenses/by/4.0/This is an open-access article distributed under the terms of the Creative Commons Attribution license, which permits unrestricted use, distribution, and reproduction in any medium, provided the original work is properly cited*):

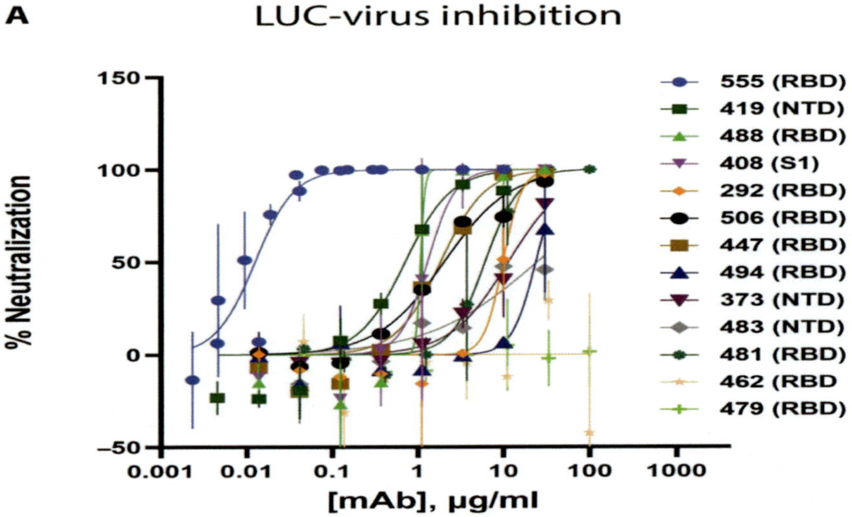

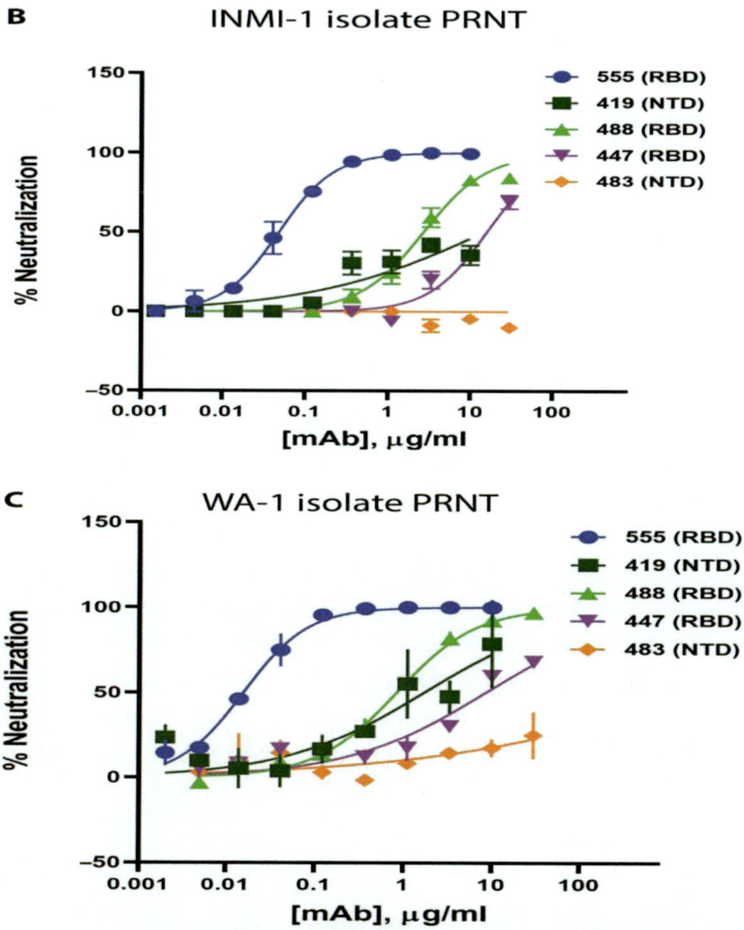
B
INMI-1 isolate PRNT
150
100
50
0
−50
% Neutralization
0.001
0.01
0.1
1
10
100
[mAb], μg/ml
555 (RBD)
419 (NTD)
488 (RBD)
447 (RBD)
483 (NTD)
C
WA-1 isolate PRNT
150
100
50
0
−50
% Neutralization
0.001
0.01
0.1
1
10
100
[mAb], μg/ml
555 (RBD)
419 (NTD)
488 (RBD)
447 (RBD)
483 (NTD)

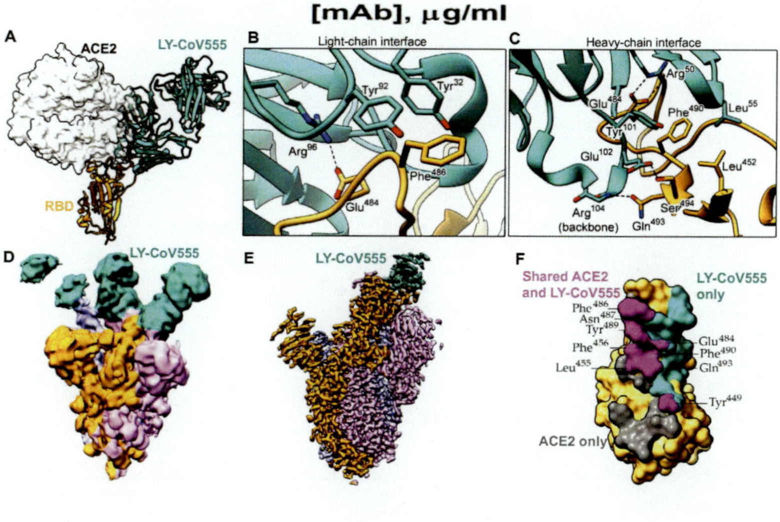
A
ACE2
LY-CoV555
RBD
B
Light-chain interface
Tyr92
Tyr32
Arg96
Phe486
Glu484
C
Heavy-chain interface
Arg50
Glu484
Phe490
Leu55
Tyr101
Glu102
Leu452
Arg104
(backbone)
Gln493
Ser494
D
LY-CoV555
E
LY-CoV555
F
Shared ACE2
and LY-CoV555
LY-CoV555
only
Phe486
Asn487
Tyr489
Phe456
Leu455
Glu484
Phe490
Gln493
Tyr449
ACE2 only

Pretreating rhesus macaques 24 hours prior to virus challenge with LY-CoV555 (control IgG1 antibody) revealed lower viral replication with lowered subgenomic RNA (sgRNA) and viral genomes (gRNA) in bronchoalveolar lavage fluid (BALF): the viral replication and load were lowered by 10^2 to 10^5 copies per milliliter vs. the control with undetected viral replication at day 3 in LY-CoV555-treated animals. Protection was also seen in the throat and the upper respiratory tract.

Figure 6 below illustrates the lowered viral replication and load in macaque ***TOP***: lower respiratory tract and ***BOTTOM***: upper respiratory tract administered a single dose of LY-CoV555 (1, 2.5, 15, or 50 mg/kg) 24h before challenge with SARS-CoV-2; qRT-PCR for (A) sgRNA (viral replication) (B) gRNA (viral load) over six days (C) sgRNA (viral replication) and (D) gRNA (viral load) on day 6. The lower limit of quantification (LLOQ) = 50 copies for genomic or subgenomic mRNA (Bryan E. Jones et al, 2021: *Copyright © 2021, Copyright © 2021 The Authors, some rights reserved; exclusive licensee American Association for the Advancement of Science. No claim to original U.S. Government Works. Distributed under a Creative Commons Attribution License 4.0 (CC BY).. https://creativecommons.org/licenses/by/4.0/This is an open-access article distributed under the terms of the Creative Commons Attribution license, which permits unrestricted use, distribution, and reproduction in any medium, provided the original work is properly cited*):

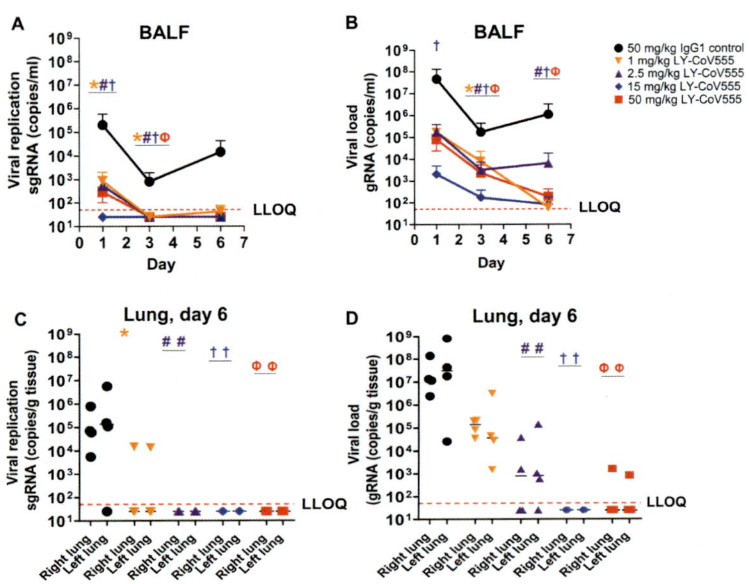

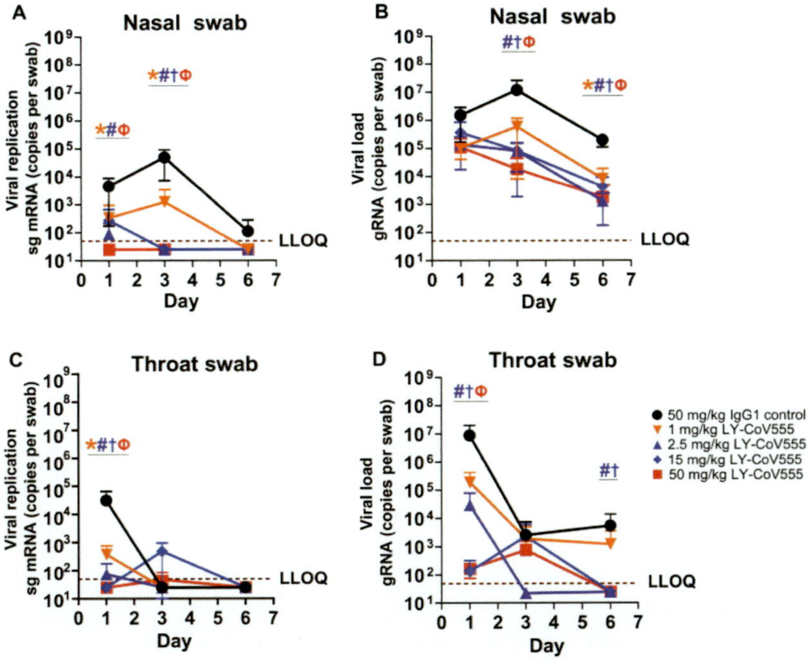

According to the authors "*LY-CoV555 was developed as a therapeutic antibody specifically to treat COVID-19. The treatment quickly entered clinical testing and demonstrated clinical efficacy and gain Emergency Use Authorization. LY-CoV555 is presently under clinical evaluation for the treatment and prevention of COVID-19* ***(NCT04411628, NCT04427501, NCT04497987, NCT04501978, NCT04518410****, and* ***NCT04634409****) in various clinical settings*".

2.5 MABS FOR ASTHMA

With a global occurrence of 300 million people, asthma occupies the 16th position among the leading causes of years lived with a disability making it a crucial non-communicable disease. The occurrence of this complex multifactorial disorder with the involvement of genetics, host factors and environmental impact can be at any time throughout life. Novel systems for personalized care are yet to be realized completely for this condition (Dharmage SC, Perret JL and Custovic A, 2019).

A slew of "triggers" have been summarized below in figure 7 (© 2019 Dharmage, Perret and Custovic. *This is an open-access article distributed under the terms of the* ***Creative Commons Attribution License (CC BY)****. The use, distribution or reproduction in other forums is permitted, provided* *the original author(s) and the copyright owner(s) are credited and that the original publication in this journal is cited, in accordance with accepted academic practice. No use, distribution or reproduction is permitted which does not comply with these terms*):

Environmental exposure	Feature	Children	Adolescents	Adults
ALLERGIC				
Airborne triggers - House dust mite (HDM)	HDM aeroallergen is a perennial asthma trigger linked to asthma incidence in high-risk children	(94–96)	(94)	(94)
- Animal hair and dander - Pollen exposure	Association between pet allergen exposure and asthma is conflicting and non-conclusive	(94, 97, 98)	(98)	(98)
- Mold (fungal) spores	Grass pollen triggers asthma exacerbations requiring emergency department attendances	(99)	(99)	
- Thunderstorm asthma	Indoor fungal spore exposure can worsen asthma control; decreased visible indoor mold reduces symptoms but not PEFM variability	(100, 101)	(100, 101)	(100, 101)
	High outdoor Alternaria exposure may contribute to severe asthma/ respiratory arrest (ages 11–25)	(102)	(103)	(103)
	Fungal spores, especially Cladosporium are associated with asthma hospitalization (ages 2–17)	(104)	(104)	
	IgE sensitization to mycoses linked to neutrophilic airway inflammation and lower lung function			(105)
	Thunderstorm asthma can be triggered by outdoor pollen, and possibly fungal, spores	(106, 107)	(106, 107)	(106, 107)
Food allergens ($n = 170$), e.g.,	In asthma, c/w non-atopy, odds for current asthma increased 3.8-fold if food allergy was likely	(108)	(108)	(108)
- Egg white - Peanut	Food allergy is an uncommon trigger in asthma, but may present as life-threatening asthma, especially to peanut and other tree nut allergens	(109, 110)		
- Tree nuts - Shellfish - Cows milk	Co-existing, poorly-controlled asthma is a risk factor for severe or fatal food-induced anaphylaxis	(110, 111)		
Occupational sensitizing agents (with latency)	Extensive lists of occupational asthmagens known to cause new-onset occupational asthma and/or exacerbate pre-existing asthma† - Sensitizing HMW agents (e.g., plant allergens like flour, flowers, latex; animal allergens by animal handlers and lab workers; biological enzymes; fungi-yeasts) - Sensitizing LMW agents (e.g., chemicals like isocyanates, reactive dyes, industrial cleaning/sterilizing agents; metals; pharmaceuticals like antibiotics, opiates; solder flux; wood dusts)		(112–115)	(112–115)
NON-ALLERGIC				
Non-allergic triggers - Respiratory viral infections - Cold air - Humidity - Exercise	Typically trigger variations in asthma symptoms and airflow limitation, and their presence increases the probability that the individual has asthma	(94)	(94)	(94)
Tobacco smoke exposure - Parental smoking	Parental smoking linked to increased incidence of childhood asthma and wheeze	(116)		
- Second-hand smoke exposure	Incident asthma from regular smoking from late childhood; may be greater for those non-allergic compared with allergic, and for those exposed to maternal smoking *in utero*	(117, 118)	(117, 118)	
- Personal smoking	Personal smoking may worsen asthma control/ exacerbations		(119)	(119)
	Personal smoking predisposes to post-BD airflow obstruction and asthma-COPD overlap			(120)
Traffic-related air pollution (TrAP)	Air pollutants (O_3, $PM_{2.5}$) linked to new asthma cases and increased ED admissions globally	(121)	(121)	(121)
- Car exhaust fumes	Long-term exposure to PM_1 may worsen asthma, especially for young males with childhood allergy	(122)	(122)	
	Pollution-related decline in FEV_1 growth was similar between those with and without asthma	(123, 124)	(123, 124)	
	Residential TrAP may contribute to new-onset and persistence of asthma in middle-aged adults			(125)
	Natural experiment of diesel exhaust exposure and adverse short-term changes in spirometry			(126)
Household air pollution (HAP)	Non-polluting home heating improved asthma symptoms, days off school, and healthcare utilization	(127)		
Occupational agents (no latency)	Airway irritants (e.g., chlorine, ammonia) - Irritant occupational asthma - Work exacerbated (pre-existing) asthma		(113, 114)	(113, 114)

BD, bronchodilator; COPD, chronic obstructive pulmonary disease; HMW, high molecular weight; FEV_1, forced expiratory volume in 1 second; LMW, low molecular weight; O_3, ozone; PM, particulate matter.

†A web-based list of asthmagens can be found at www.occupationalasthma.com.

Inhaled corticosteroids (ICS) document favorable responses in patients with type 2 asthma (manifested as eosinophilic inflammation) yet, some individuals necessitate high doses or manifest symptoms despite treatment. Several mAbs targeting vital inflammatory molecules have been designed as reviewed by Edris et al (2019) in *Respiratory Research.* As IL-5 is vitally involved in eosinophil biology, it has received the limelight with mAbs mepolizumab, benralizumab and reslizumab having received approval from FDA and EMA. Mepolizumab binds to soluble interleukin (IL)-5; Reslizumab causes downregulation of the pathway; benralizumab binds to the interleukin-5 receptor α (IL-5Rα) and NK cell (natural killer) FcγIIIRa receptors to facilitate apoptosis of eosinophils. Another target is IL-13 associated with smooth muscle contraction and B-cell switch to the allergic IgE. Molecules examined in trials are inclusive of Lebrikizumab and GSK679586 bind to IL-13; Tralokinumab inhibits downstream pathways. Hospital-associated visits were lowered by Dupilumab that binds to IL-4 and has received FDA approval as an add-on maintenance therapy in moderate to severe asthma. Other molecules being probed or in trials are MEDI-528 targeting IL-9 a, Tezepelumab binds thymic stromal lymphopoietin (TSLP) and Daclizumab binds to the IL-2R α chain (CD25).

Figure 8 below summarizes the mAbs approved and their targets in type-2 asthma therapy: **a**) Mepolizumab and reslizumab bind to IL-5; Benralizumab is an IL-5 receptor blocker **b)** Tezepelumab inhibits the inflammatory cytokine TSLP; Dupilumab binds to the alpha subunit of IL-4 receptor; Lebrikizumab binds to IL-13 cytokines and tralokinumab binds to IL-13 receptor on B cells hampering IgE production(Edris et al, 2019: *Open Access This article is distributed under the terms of the Creative Commons Attribution 4.0 International License (http://creativecommons.org/licenses/by/4.0/), which permits unrestricted use, distribution, and reproduction in any medium, provided you give appropriate credit to the original author(s) and the source, provide a link to the Creative Commons license, and indicate if changes were made. The Creative Commons Public Domain Dedication waiver (http://creativecommons.org/publicdomain/zero/1.0/) applies to the data made available in this article, unless otherwise stated*):

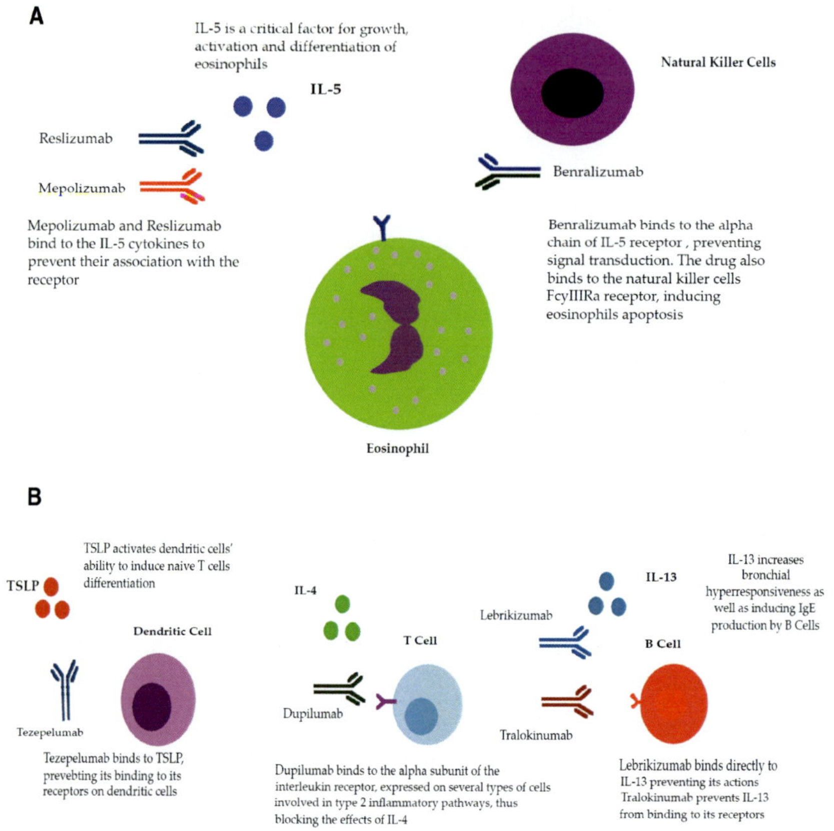

The first mAb to receive FDA approval was Omalizumab back in 2003 with its use documented in moderate to severe asthma entailing frequent exacerbations in spite of high-dose inhaled steroids. Given that the Global Initiative of Asthma (GINA) recommends mAb use prior to oral corticosteroids (OCs), omalizumab has been advised in IgE-mediated asthma. With this background, a case study was published in Frontiers in Pharmacology by Domingo, Pomares , Morón and Sogo (2020) in a 55-year-old female with a past history of chronic OC treatment. Administration of 300 mg/month sub-cutaneous (sc) omalizumab was recommended to which a mild improvement was reported after 16 weeks. There were no severe asthma exacerbations requiring hospital admission. Subsequently, a switch to mepolizumab was suggested upon its release to address her elevated IgE levels. Allergy reappeared following 100 mg of sc mepolizumab for 14 months that led the team to propose a dual therapy encompassing both mAbs concomitantly on the same day (omalizumab 300 mg/month plus mepolizumab 100 mg per month, both sc).

The team reported a notable improvement following 24 months of dual therapy with normal FEV_1 (Forced expiratory volume in the first second that was indicative of obstruction in the patient) of 83% of the predicted value in the first year and 96% (normal range) at the second year from her earlier 38% of the predicted figure. Further, the dose of OC was also lowered to 2.5 mg per day of prednisolone. The treatment reported no side effects. The team suggested the efficacy of a combination of the two mAbs in some severe allergic asthma patients with persistently high eosinophil levels.

Figure 9 below is illustrative of dual mAb therapy in the severe asthma patient (A) FEV1 changes from only omalizumab to only mepolizumab and in the third both mepolizumab and omalizumab. The dual mAb therapy documented a conspicuous improvement in the FEV1 (B) Lowered oral corticosteroid (OC) administration (mean monthly accumulated intake) for omalizumab, the 14 months of mepolizumab in the 24 months of omalizumab plus mepolizumab (© 2020 Domingo, Pomares, Morón and Sogo. *This is an open-access article distributed under the terms of the* ***Creative Commons Attribution License (CC BY)****. The use, distribution or reproduction in other forums is permitted, provided the original author(s) and the copyright owner(s) are credited and that the original publication in this journal is cited, in accordance with accepted academic practice. No use, distribution or reproduction is permitted which does not comply with these terms*):

A

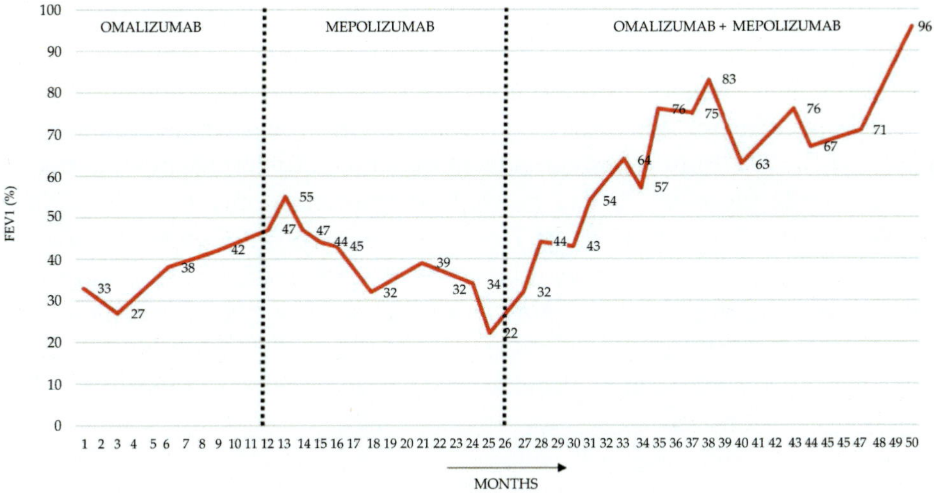

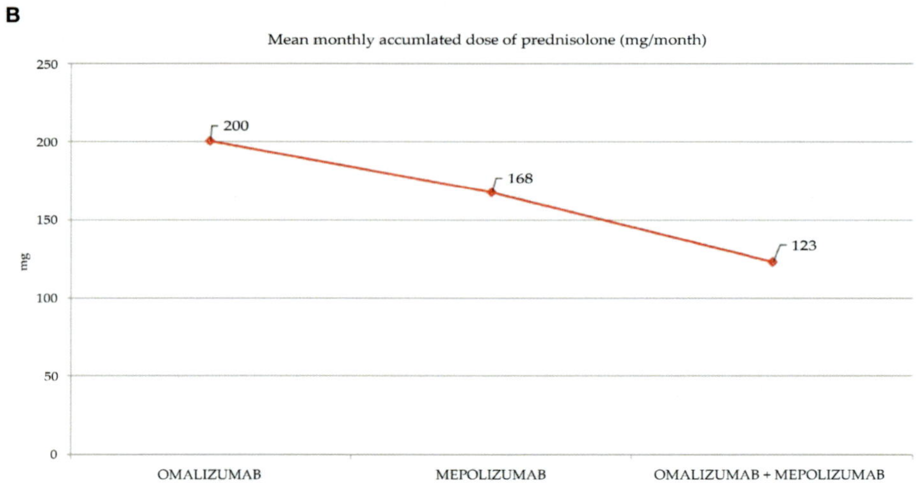

2.6 THE HOPE FOR HIV

The peril of human immunodeficiency virus (HIV) looms even now with an average loss of 36.3 million lives so far as per the WHO. The year 2020 saw 680000 [480 000–1.0 million] people dying of HIV-related causes while 1.5 million [1.0–2.0 million] individuals were infected. A slew of approaches can be envisaged here with targeting the CD4-binding site on HIV gp120 to directly block virus attachment to target cells or targeting the prefusion stages. Figure 10 below is illustrative of mAbs to target HIV: In **A** the virus can be targeted by binding to specific virus surface proteins (1) binding to the viral receptor or co-receptor on the host (2) interfering with conformational changes for viral core release (3) targeting endocytosis (4) Fc-mediated effector functions (5) preventing virion release (6) Augmented phagocytosis (7). In **B** mAbs can be employed to activate receptors to boost immune cells (agonist) or can block host immunosuppressive molecules (antagonism) (Flego et al, 2013: *This article is published under license to BioMed Central Ltd. This is an Open Access article distributed under the terms of the Creative Commons Attribution License (http://creativecommons.org/licenses/by/2.0), which permits unrestricted use, distribution, and reproduction in any medium, provided the original work is properly cited*):

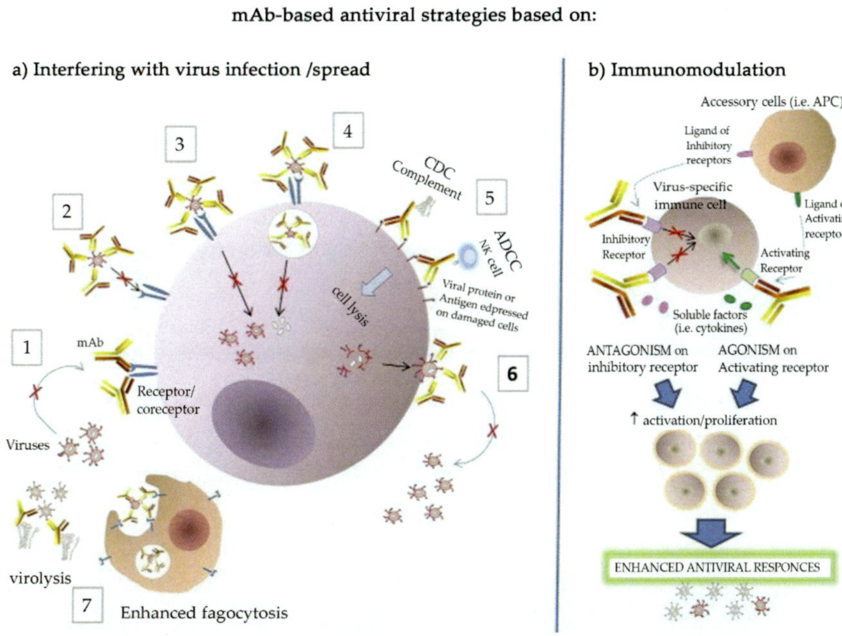

Despite the use of the only FDA-approved treatment modality for HIV/AIDS as of 2021, antiretroviral therapy (ART) aims at maintaining low viral loads in the system. The quest for mAb-based therapies initially identified neutralizing antibodies (NAbs) that were cloned and probed from infected patients. Early work in the 1980s–1990s unveiled the development of infection with virus-neutralizing polyclonal IgG transfer from HIV subjects. First-generation bNAbs conferred non-human primate (NHP) models protection from simian immunodeficiency virus (SHIV). Subsequent work on B-cells developed second-generation bNAbs that were potent and could protect NHP models and humanized mice. Ongoing work is exploring trials in humans and new systems for delivery and molecule cocktails (Jaworski, Vendrell and Chiavenna, 2017; Spencer, Shapiro, Haigwood and Hessell, 2021).

Figure 11 below is illustrative of mAb-based work for targeting HIV (© 2021 Spencer, Shapiro, Haigwood and Hessell. *This is an open-access article distributed under the terms of the* ***Creative Commons Attribution License (CC BY)****. The use, distribution or reproduction in other forums is permitted, provided the original author(s) and the copyright owner(s) are credited and that the original publication in this journal is cited, in accordance with accepted academic practice. No use, distribution or reproduction is permitted which does not comply with these terms*):

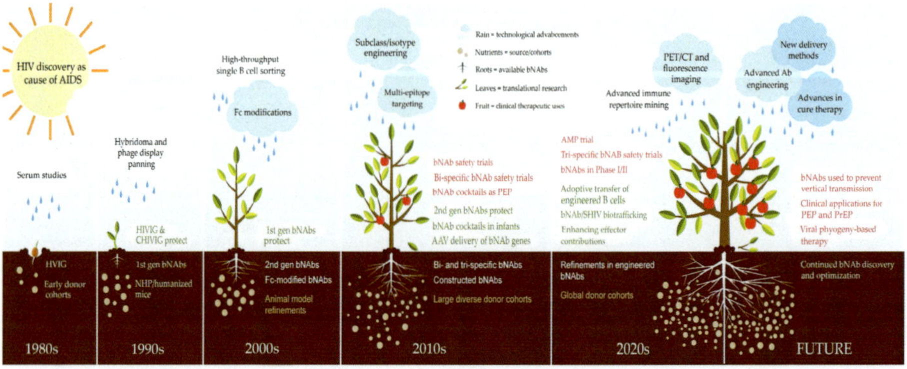

Table 7 below documents few first and second generations of anti-HIV-1 neutralizing monoclonal antibodies (NmAbs) (© 2017 Jaworski, Vendrell and Chiavenna. *This is an open-access article distributed under the terms of the **Creative Commons Attribution License (CC BY)**. The use, distribution or reproduction in other forums is permitted, provided the original author(s) or licensor are credited and that the original publication in this journal is cited, in accordance with accepted academic practice. No use, distribution or reproduction is permitted which does not comply with these terms*):

Env subunit	Epitope region	NmAb name	Potency (*in vitro*)[a]	Breadth (*in vitro*)[b]	Protection efficacy (*in vivo*)[c]
Gp120	CD4-binding site (conformational epitope)	b12[d] (18)	1.8 (47); 2.82 (45)	17 (47); 10 (45, 46)	25 mg/kg (4/4); 5 mg/kg (2/4) (28, 29, 34)
		VRC01 (01–03) (47)	0.33 (45, 47); 0.22 (61); 0.3 (55); 0.9 (44)	72 (47); 74 (45); 77 (61); 75 (55)	5 mg/kg (6/6); 0.3 mg/kg (5/12) (60)
		VRC07 (61)	0.11 (61); 0.16 (103)	83 (61, 103)	0.2 mg/kg (3/4); 0.05 (0/4) (61)
		3BNC117 (55, 60, 117) (44)	0.134 (103); 0.11 (55)	79 (103); 77 (55)	5 mg/kg (3/4); 1 mg/kg (1/4) (62)
		NIH45-46 (44)	0.2 (55); 0.41 (44)	76 (55)	20 mg/kg (0/4) (62)
		G54W	0.04 (141)	NC	ND
		12A12 (44)	NC	NC	ND
		VRCPG04 (48, 142)	0.196 (48); 0.2 (45)	64 (48); 65 (45)	ND
		VRC-CH31 (30–34) (48)	0.098 (48)	70 (48)	ND
		HJ16 (54)	ND	ND	ND
		N6	0.038 (102)	96 (102)	ND
	V3 loop (lineal epitope)	447-52D[d] (19)	NC	NC	ND
		HGN194 (54)	NC	NC	ND
	V1/V2-glycan site (quaternary epitope linked to Asn160)	PG9 (46)	0.23 (45, 46); 0.142 (103); 0.2 (55)	57 (46); 54 (45); 73 (103); 65 (55)	20 mg/kg (5/6); 5 mg/kg (3/6) (60)
		PG16 (46)	0.15 (46); 0.15 (55)	51 (46); 59 (55)	ND
		CH01-04 (143)	NC	40 (143)	ND
		PGT145 (141–145) (45)	0.2 (45)	52 (45)	ND
	V3-glycan supersite (conformational epitope linked to Asn 332)	2G12[d] (21, 144)	2.38 (45, 46)	11 (45)	40 mg/kg (3/5) (27, 30, 33)
		PGT121 (121–123) (45, 120, 145)	0.03 (45)	57 (45)	1 mg/kg (5/5): 0.2 mg/kg (3/5) (59)
		PGT128 (125–128) (45, 125)	0.02 (45); 0.096 (103)	60 (45); 56 (103)	10 mg/kg (5/5); 2 mg/kg (2/5) (146)
		10-1074 (45, 56)	0.4 (103)	54 (103)	5 mg/kg (4/4) 1 mg/kg (1/4) (62)
		PGT130-131 (45)	0.16–0.52 (45)	ND	ND
		PGT133-134 (45)	ND	ND	ND
		PGT135 (135–137) (45, 121)	0.17 (45)	23 (45)	ND
Gp41	Gp41 MPER (lineal epitope)	4E10[d] (17)	3.41 (45, 46); 1.93 (55)	13 (45); 37 (55)	50 mg/kg (6/6) (31)
		2F5[d] (20)	2.30 (46); 14.6 (55)	19 (46); 16 (55)	50 mg/kg (6/6) (27, 31, 33)
		Z13 (26)	ND	ND	ND
		10E8 (55)	0.389 (103); 0.35 (55)	74 (103); 72 (55)	5 mg/kg (6/6); 0.3 mg/kg (3/6) (60)
		HK20 (54)	ND	ND	ND
Gp41-gp120 interfase	N-linked glycans adjacent to CD4bs and gp41	8ANC195 (44, 127)	ND	ND	ND
	Quaternary structure of pre-fused gp41	35O22 (119)	ND	ND	ND
	Quaternary structure of pre-fused and cleaved gp41	PGT151 (151–158) (116, 118)	ND	ND	ND
	Fusion peptide (gp41) and glycan at Asn88 (gp120)	N123-VRC34.01 (57)	ND	ND	ND
	Quaternary structure of pre-fused gp41 and a glycan at Asn88 (gp120)	3BC176, 3BC315 (49, 147)	1.69–10 (49, 147)	ND	ND

[a]*Median 50% inhibitory concentration (IC50) (μg/ml). Results were obtained from TZM-bl neutralization assay. Panels of 100–200 pseudoviruses, representative of all HIV-1 clades, were used to measure potency.*

[b]*Percentage of virus neutralized with an IC50 < 1 μg/ml. Results were obtained from TZM-bl neutralization assay. Panels of 100–200 pseudoviruses, representative of all HIV-1 clades, were used to measure breadth.*

[c]*Passive studies performed in non-human primates (NHP): indicated is the proportion of protected animals at each Ab dose.*

[d]*First-generation NmAbs.*

ND, no data are available; NC, data available cannot be compared.

Clinically, pharmacokinetic (PK) probing has been pursued scrutinizing the tolerability, efficacy and safety of bnAbs. For instance, a bnAb called VRC07 targeting CD4bs has been tested in healthy uninfected individuals [VRC 605 (NCT03015181)]: the study at the National Institutes of Health Clinical Center in Bethesda, MD, USA entailed the administration of a single administration of 1 mg/kg to 40 mg/kg IV and three administrations (every 90 days) of 5 mg/kg SC and 20 mg/kg IV of VRC07-523LS that was reported as well tolerated with serious adverse events or dose-limiting toxic effects. According to the researchers "*VRC07-523LS is a strong and practical candidate for inclusion in HIV-1 prevention and therapeutic strategies. The results from this trial also indicate that an HIV-1 broadly neutralising monoclonal antibody engineered for improved pharmacokinetic and neutralisation properties can be safe for clinical use*" (Gaudinski et al, 2019).

A phase-I trial in 2017 by Mayer and team (Clinical Trials Registration: NCT02165267) assessed VRC01 targeting HIV-1 CD4. The study entailed the enrolment of low-risk, uninfected healthy participants (n=88) between September 9, 2014, and July 15, 2015, enrolled in 6 clinical research sites (United States) affiliated with the HVTN (HIV Vaccine Trials Network). Random allotment of the volunteers (18–50 years old) of diverse origins into the following regimens: a loading dose of 40 mg/kg IV of VRC01 followed by five 20 mg/kg IV VRC01 doses every 4 weeks (treatment group 1 [T1], $n = 20$); eleven 5 mg/kg subcutaneous (SC) VRC01 (treatment group 3 [T3], $n = 20$); placebo (placebo group 3 [P3], $n = 4$) doses every 2 weeks; or three 40 mg/kg IV VRC01 doses every 8 weeks (treatment group 2 [T2], $n = 20$). Three 10 or 30 mg/kg IV VRC01 doses were administered to Treatment groups T4 and T5 ($n = 12$ each) every 8 weeks, respectively. The flow of the work (A) and specimen collection (B) are depicted below in figure 12 (Mayer et al, 2017: *This is an open access article, free of all copyright, and may be freely reproduced, distributed, transmitted, modified, built upon, or otherwise used by anyone for any lawful purpose. The work is made available under the Creative Commons CC0* public domain dedication):

Key: ADCC, antibody-dependent cellular cytotoxicity; ADCP, antibody-dependent cellular phagocytosis; CONSORT, Consolidated Standards of Reporting Trial; IV, intravenous; P3, placebo group 3; q, quodque; SC, subcutaneous; T1, treatment group 1; T2, treatment group 2; T3, treatment group 3; T4, treatment group 4; T5, treatment group 5

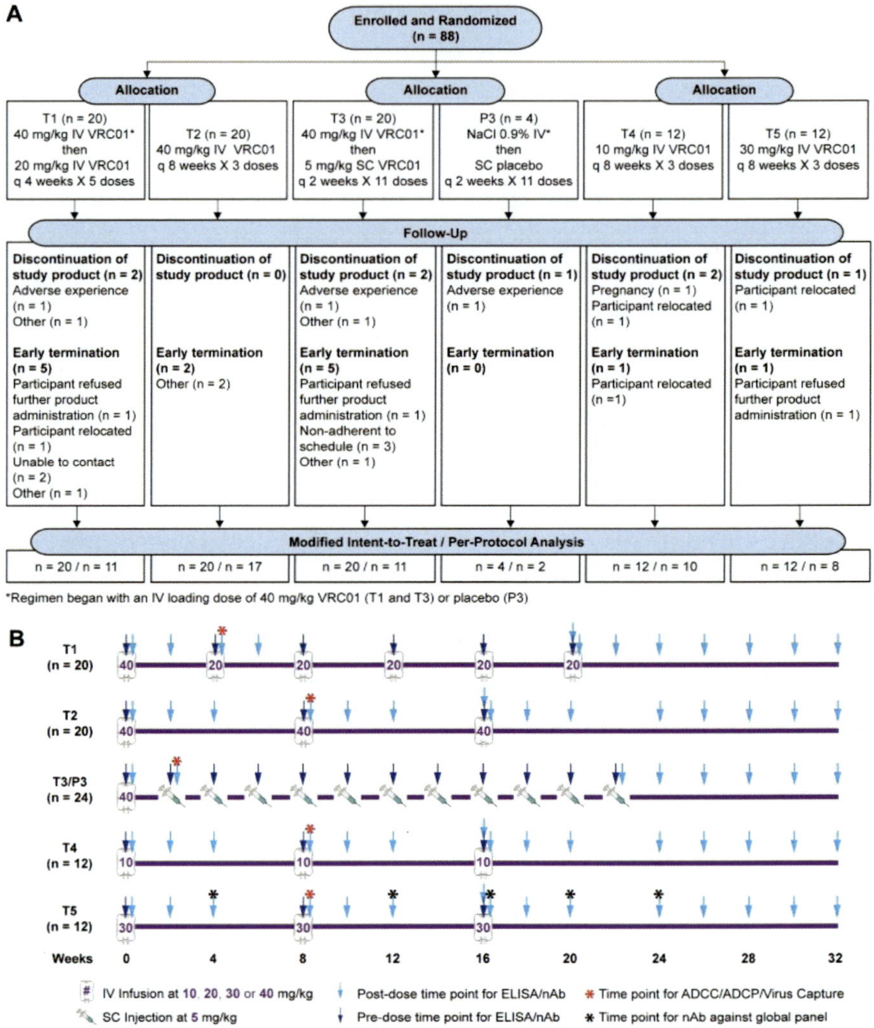

Follow-up was for 32 weeks after the first VRC01 dose with no anti-VRC01 antibodies or dose-limiting toxicities or serious adverse events across 249 IV infusions and 208 SC injections done. The functional aspects of VRC01 in the serum were retained (in terms of virion capture, neutralization, phagocytosis and antibody-dependent cellular cytotoxicity). Overall, the researchers concluded that the rare, transient and mild adverse events coupled with trough data that revealed neutralization of a slew of strains *in vitro* (50% inhibitory concentration [IC50] > 5 μg/ml) pave the way for further scrutiny and employment of VRC01 for prophylaxis.

Figure 13 below documents the outcome of the phase-I trial probing the promise of VRC01 for HIV prophylaxis:

TOP: systemic (A) and local (B) reactions based on DAIDS(Division of AIDS) Table 8 for Grading the Severity of Adult and Pediatric Adverse Events, Version 1.0, December 2004; MITT, modified intent-to-treat; P3, placebo group 3; q, quodque; SC, subcutaneous; T1, treatment group 1; T2, treatment group 2; T3, treatment group 3; T4, treatment group 4; T5, treatment group 5 (Mayer et al, 2017: *This is an open access article, free of all copyright, and may be freely reproduced, distributed, transmitted, modified, built upon, or otherwise used by anyone for any lawful purpose. The work is made available under the Creative Commons CC0* public domain dedication):

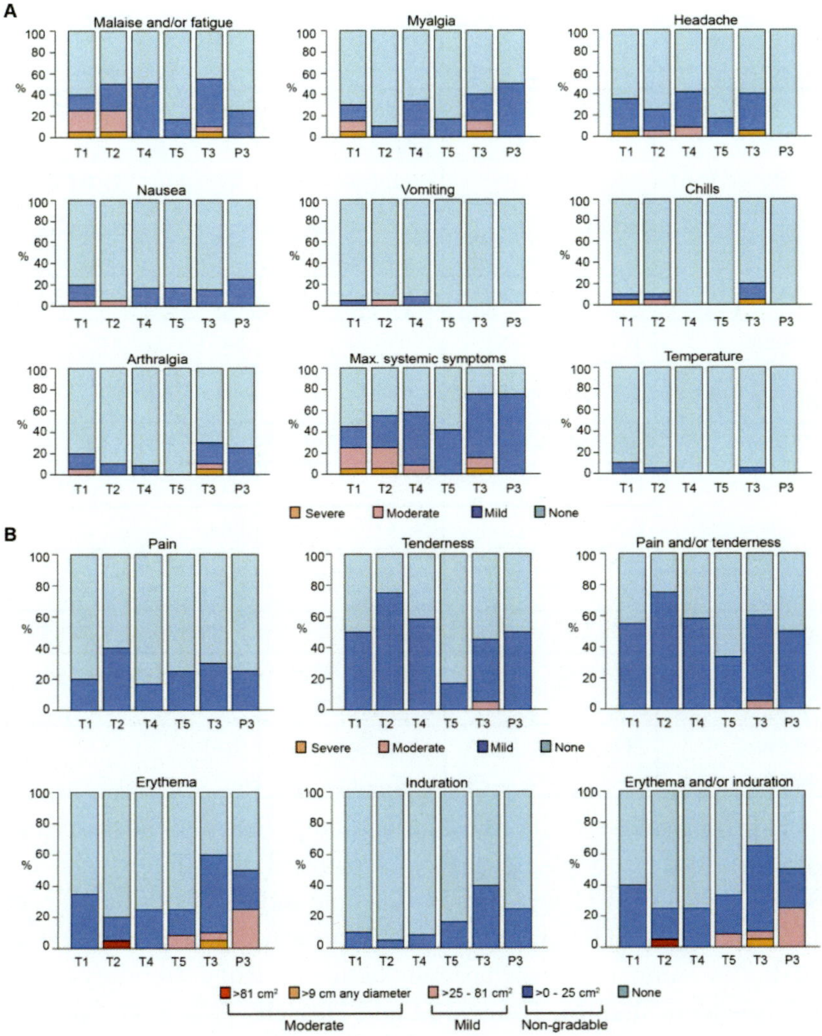

Several such studies are depicted in table 9 below ():

Key: [1]AAV delivery system; [2]tested in HIV-infected participants; NA, Not available

Epitope region targeted	**bnAb**	**Half-life**	**Reference**
CD4-binding site (CD4-BS)	VRC01	15 days; 71 days (-LS)	NCT02165267 (73, 74);
	VRC07-523	40 days (-LS)	NCT03015181 (75, 76);
	3BNC117	18 days; 61 days (-LS)	NCT02018510 (77, 78);
	N6	>30 days (-LS)	NCT03538626 (79);
Membrane-proximal external region (MPER)	10E8	N/A	NCT03565315 (80);
V1V2 loop glycan	PGDM1400	N/A	NCT03205917
	PG9[1]	N/A	NCT01937455 (81)
	CAP256-VRC26.25	N/A	PAC-TR202003767867253 (82);
V3 loop glycan	PGT121	22 days	NCT02960581 (83);
	10-1074	24 days; 73.5 days (-LS)	NCT02511990 (78, 84);
gp120 outer domain	2G12	16.5 days[2]	(85)

Apart from treatment, the role of this therapy has also been scrutinized for post-exposure prophylaxis (PEP). Treatment of infants with bNAbs VRC07-523 and PGT121 in a rhesus macaque model of perinatal infection revealed treating of SHIV infection while preventing reservoir formation. Table 10 below summarizes such prophylactic usage of antibodies in animals (

NHP model	bNAb	Target	Dose	Treatment day post-challenge	Outcome	References
$SHIV_{SF162P3}$, intravaginal (adults)	PGT121	V3-glycan	2 mg/kg (1×)	−1	12/12 (100%) protected from durable infection; tissue virus detected at day 1–3 in in three NHPs was cleared by days 7–10	(75)
$SHIV_{SF162P3}$, oral (infants)	PGT121 and VRC07-523	V3 glycan and CD4bs	5 mg/kg each (4×) 40 mg/kg total	1, 4, 7, 10	4/4 (100%) aviremic for 6 months	(84)
			20 mg/kg each (4×) 160 mg/kg total	1, 4, 7, 10	6/6 (100%) aviremic for 6 months (with serial necropsy of additional NHPs showing tissue virus detected at day 2 was cleared by day 7–14)	
$SHIV_{SF162P3}$, oral (infants)	PGT121 and VRC07-523	V3 glycan and CD4bs	20 mg/kg each (1×) 40 mg/kg total	1.25	6/6 (100) aviremic for 8 month	(85)
			5 mg/kg each (4×) 40 mg/kg total	2, 4, 7, 10	3/6 (50%) aviremic or transient blips over 10 months	
$SHIV_{AD8EO}$, intrarectal, or intravenous (adults)	10–1,074 and 3BNC117	V3 glycan and CD4bs	10 mg/kg each (3×) 60 mg/kg total	3, 10, 17	3/13 (23%) aviremic or transient blips for 12 months	(86)
$SHIV_{SF162P3}$, intravenous (adults)	PGT121 and VRC07-523	V3 glycan and CD4bs	20 mg/kg each (1×) 40 mg/kg total	10 (followed by ART on days 21–112)	0/6 (0%) aviremic–all rebounded after ART	(87)

2.7 GLYCOENGINEERING MABS

For therapy, a malady such as an oncogenic malignancy can be targeted by a naked antibody that is aimed at immune checkpoint inhibition or the microenvironment or inducing apoptosis or complement-mediated cytotoxicity (CDC) or antibody-dependent cellular cytotoxicity (ADCC). There are recent developments in this in terms of "glycoengineering". The predominant recombinant therapeutic mAbs are glycosylated immunoglobulin G (IgG) molecules containing glycans attached to the amide nitrogen atom of asparagine 297 (N297) in the crystallizable fragment (Fc) region. The Fc glycosylation has received the limelight in terms of how mAb stability, CDC, ADCC and safety are affected by glycoforms. For instance, when the core Fuc residue in the N-glycans is absent, the ADCC was augmented by a factor of 100 in this afucosylated anti-HER2 IgG vs. the one containing the Fuc residue. Defucosylated antibodies also possess augmented binding to the low-affinity allotype of FcγRIIIa-158F to hence broaden their use irrespective of polymorphisms. This has led to several approaches of glycoengineering where defucosylated mAb production is being probed (Li, Zhu, Chen, Feng and Dimitrov, 2017; Lu et al, 2020).

Figure 14 below shows the crystal structures of Fc–FcγRIIIa with defucosylated IgGs documenting several bonds that facilitate the defucosylated Fc to bind to FcγRIIIa with many of these non-covalent bonds hindered by fucose in the fucosylated Fc setting the stage for defucosylated antibodies. The non-fucosylated crystallizable fragment (Fc) (PDB ID: 3SGK) and fucosylated Fc (PDB ID: 3SGJ) complexed with FcγRIIIa with a high mannose glycan on N162 are depicted (A) non-fucosylated Fc–FcγRIIIa complex produced by PyMOL. The oligosaccharides and part of hydrogen bonding formation residues are depicted as spheres and sticks. The hydrogen bonds are depicted as black dash lines. (B) fucosylated Fc–FcγRIIIa complex. The core fucose is the dot representation (© 2017 Li, Zhu, Chen, Feng and Dimitrov: *This is an open-access article distributed under the terms of the Creative Commons Attribution License (CC BY). The use, distribution or reproduction in other forums is permitted, provided the original author(s) or licensor are credited and that the original publication in this journal is cited, in accordance with accepted academic practice. No use, distribution or reproduction is permitted which does not comply with these terms*):

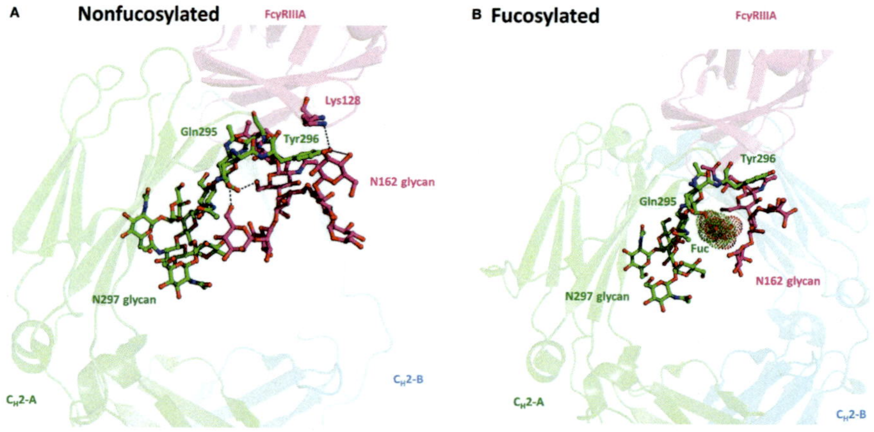

Various glycoengineering research approaches for defucosylated antibody production are summarized in table 11 below (© 2017 Li, Zhu, Chen, Feng and Dimitrov: *This is an open-access article distributed under the terms of the Creative Commons Attribution License (CC BY). The use, distribution or reproduction in other forums is permitted, provided the original author(s) or licensor are credited and that the original publication in this journal is cited, in accordance with accepted academic practice. No use, distribution or reproduction is permitted which does not comply with these terms*):

Key: NA, not applied; RhD, rhesus D antigen; GFT, GDP-fucose transporter; GM2, ganglioside mono-sialic acid 2; GD3, ganglioside di-sialic acid 3; GMD, GDP-mannose 4,6-dehydratase; GnT III, GlcNAc transferase III

Cell type			**Glycoengineering modification**	**Company and technology platform**	**Antibody name**	**Targets**	**Development status**	**Reference**
Non-mammalian cells	Yeasts		och or alg3 KO	NA GS4.0	Rituximab	Anti-CD20	NA	(45–47)
	Plants		RNAi of β-1,2-XylT and α-1,3-FucT	MAPP Biopharmaceutical	ZMAPP	Anti-Ebola	Phase 2/3	(48, 49)
Mammalian cells	YB2/0		Intrinsic low core fucose content YB2/0 (FUT8 low); Lec13 (GMD low)	LFB Biotechnologies EMABling Technology	Ublituximab; roledumab	Anti-CD20; anti-RhD	Phase 2/3	(50)
	CHO cells	Lec13 cells		Biowa	Hu4D5	Anti-HER2	NA	(51)
		WT CHO	Addition of sugar analog into culture medium (process engineering)	Seattle Genetics SEA Technology	SEA-CD40	Anti-CD40	Phase 1	(52)
		Genetically modified CHO	GMD KO	GMD knockout CHO/DG44	Rituximab	Anti-CD20	NA	(53)
			RMD overexpression	ProBioGen, GlymaxX Technology	Trastuzumab	Anti-HER2	NA	(54)
			GFT KO	CRISPR-Cas9 KO GFT CHO-F6	LSEVh-LS-F	Anti-HIV-1 Env	Pre-clinic	(55, 56)
			GnT III overpression	Roche GlycoMab Technology	Obinutuzumab	Anti-CD20	Approved	(57)
					RG7116	Anti-HER3	Phase 1	(58)
			RNAi of FUT8 and/or GMD	FG16	KM2160	Anti-CCR4	NA	(59, 60)

			FUT8 KO	Kyowa Hakko Kirin Potelligent Technol-ogy	Mogamuli-zumab	Anti-CCR4	Approved	(61)
					Benrali-zumab, ec-romeximab, MEDI-551, BIW-8962, KHK2804, 2823, 2898, 4083	Anti-IL-5Rα, GD3, CD19, GM2, CD123, CD98	In clinical trials	(31)
				FUT8 knockout CHO/ DG44	NA	Anti-CD20	NA	(62)
	293 Free-Style cells		α-Mannosidase inhibitors kifunensine	NA	4Dm2m-F	Anti-HIV-1 Env	NA	(63)

Further, therapeutic mAbs and their glycosylation schemes have been documented in table 12 below (© 2017 Li, Zhu, Chen, Feng and Dimitrov: *This is an open-access article distributed under the terms of the Creative Commons Attribution License (CC BY). The use, distribution or reproduction in other forums is permitted, provided the original author(s) or licensor are credited and that the original publication in this journal is cited, in accordance with accepted academic practice. No use, distribution or reproduction is permitted which does not comply with these terms*):

Key: CCR4, C–C chemokine receptor type 4; EGFR, epidermal growth factor receptor; ATLL, adult T-cell leukemia/lymphoma; CTCL, cutaneous T-cell lymphomas; CLL, chronic lymphocytic leukemia; NHL, non-Hodgkin's lymphoma; NA, not applicable; mAb, monoclonal antibody

mAbs name	**Target**	**Indication**	**Glycol modifi-cation**	**Develop-ment status**	**Reference**
Otelixizumab	CD3	Type I diabe-tes, rheuma-toid arthritis	Aglycosylated	Phase I (completed)	(35)
MTRX-1011A	CD4	Rheumatoid arthritis, cuta-neous lupus	Aglycosylated	Phase I (completed)	(36)
Mogamulizumab	CCR4	ATLL, CTCL	Afucosylated	Approved	(37, 38)
MDX-1342	CD19	Relapsed or refractory CLL	Afucosylated	Phase I	(39)

Obinutuzumab	CD20	CLL, follicular lymphoma	Low fucose	Approved	(40)
DI-B4	CD19	CD19-positive indolent B-cell lymphoma	Low fucose	Phase I	(41)
RG7160	EGFR	EGFR-positive solid tumors	Bisected; non-fucosylated	Phase II	(42)
GTMAB2.5GEX	Mucin 1	*A-MUC1*-positive ovarian cancer	Glycooptimized	Phase II	(43)
Rituximab	CD20	CLL and NHL	Galactosylated	NA	(44)
Intravenous immunoglobulin	NA	Autoantibody-driven inflammation	Sialylated	NA	(13)

2.8 MABS AND CANCER

The maladies being addressed by mAbs can all be traced back to cancer: early work in the 1980s identified cancerous B cells in non-Hodgkin's lymphoma expressing CD 20 that made it the first target for mAbs. The first approval was issued to rituximab: the mAb targeting this molecule. Clinically, mAbs against HER2 are used for breast cancer while EGFR is being targeted for colorectal cancer (Zahavi and Weiner, 2020).

Table 13 below documents mAbs approved by the FDA for cancer (Zahavi and Weiner, 2020: © 2020 by the authors. Licensee MDPI, Basel, Switzerland. *This article is an open access article distributed under the terms and conditions of the Creative Commons Attribution (CC BY) license: http://creativecommons.org/licenses/by/4.0/*):

Name	**Antigen**	**Format**	**Indications (Year of First Approval) based on the FDA drug database**
Unconjugated Antibodies			
Atezolizumab	PD-L1	Humanized IgG1	Bladder, Non-small cell lung (2016), and Triple-negative breast (2019) cancers (2019)
Avelumab	PD-L1	Human IgG1	Urothelial Carcinoma (2017) and Merkel Cell Carcinoma (2017)
Bevacizumab	VEGF	Humanized IgG1	Colorectal (2004), Non-small cell lung (2006), Renal (2009), Glioblastoma (2009), and Ovarian (2018) Cancers
Cemiplimab	PD-1	Human IgG4	Cutaneous squamous-cell carcinoma (2018)

Cetuximab	EGFR	Chimeric IgG1	Colorectal cancer (2004) and Head and neck squamous cell carcinoma (2006)
Daratumumab	CD38	Human IgG1	Multiple Myeloma (2015)
Dinutuximab	GD2	Chimeric IgG1	Neuroblastoma (2015)
Durvalumab	PD-L1	Human IgG1	Bladder Cancer (2017)
Elotuzumab	SLAMF7	Humanized IgG1	Multiple Myeloma (2015)
Ipilimumab	CTLA-4	Human IgG1	Melanoma (2011) and Renal cell carcinoma (2018)
Isatuximab	CD38	Chimeric IgG1	Multiple Myeloma (2020)
Mogamulizumab	CCR4	Humanized IgG1	Cutaneous T-cell lymphoma (2018)
Necitumumab	EGFR	Human IgG1	Non-small cell lung cancer (2015)
Nivolumab	PD-1	Human IgG4	Melanoma (2014), Lung (2015), and Renal (2018) cancers
Obinutuzumab	CD20	Humanized IgG2	Chronic lymphocytic leukemia (2013)
Ofatumumab	CD20	Human IgG1	Chronic lymphocytic leukemia (2014)
Olaratumab	PDGFRα	Human IgG1	Sarcoma (2016)
Panitumumab	EGFR	Human IgG2	Colorectal Cancer (2006)
Pembrolizumab	PD-1	Humanized IgG4	Melanoma (2014), Various (2015-)
Pertuzumab	HER2	Humanized IgG1	Breast cancer (2012)
Ramucirumab	VEGFR2	Human IgG1	Gastric cancer (2014)
Rituximab	CD20	Chimeric IgG1	B-Cell Lymphoma (1997)
Trastuzumab	HER2	Humanized IgG1	Breast cancer (1998)
Antibody–Drug Conjugates (ADCs)			
Gemtuzumab ozogamicin	CD33	Humanized ADC	Acute myeloid leukemia (2000)
Brentuximab vedotin	CD30	Chimeric ADC	Hodgkin's lymphoma and Anaplastic large-cell lymphoma (2011)
Trastuzumab emtansine	HER2	Humanized ADC	Breast cancer (2013)
Inotuzumab ozogamicin	CD22	Humanized ADC	Acute lymphoblastic leukemia (2017)
Polatuzumab vedotin	CD79B	Humanized ADC	B-Cell Lymphoma (2019)
Enfortumab vedotin	Nectin-4	Human ADC	Bladder cancer (2019)
Trastuzumab deruxtecan	HER2	Humanized ADC	Breast cancer (2019)

Sacituzumab govitecan	TROP2	Humanized ADC	Triple negative breast cancer (2020)
Moxetumomab pasudotox	CD22	Mouse ADC	Hairy-cell leukemia (2018)
Ibritumomab tiuxetan	CD20	Mouse IgG1-Y90 or In111	Non-Hodgkin's lymphoma (2002)
Iodine (I131) tositumomab	CD20	Mouse IgG2-I131	Non-Hodgkin's lymphoma (2003)
Blinatumomab	CD19, CD3	Mouse BiTE	Acute lymphoblastic leukemia (2014)

2.9 FUTURE AND FINAL POINTS

The development of mAbs for therapy entails the following as depicted in figure 15 below: **a** The traditional mouse hybridoma approach entailing mice immunization with desired antigens to trigger an immune response. The fusion of harvested splenocytes with myeloma cells produces hybridoma cells to secrete antibodies for a long time. Screening for selected leads facilitates chimeric or humanized antibody generation. **b** Phage display is inclusive of selection of positive clones of a human phage-displayed human antibody library. **c** Transgenic system **d** The single B cell technique is based on human mAb production from flow cytometric isolation of B-cells from PBMCs of donors (Lu et al, 2020: *Open Access: This article is distributed under the terms of the Creative Commons Attribution 4.0 International License (http://creativecommons.org/licenses/by/4.0/), which permits unrestricted use, distribution, and reproduction in any medium, provided you give appropriate credit to the original author(s) and the source, provide a link to the Creative Commons license, and indicate if changes were made. The Creative Commons Public Domain Dedication waiver (http://creativecommons.org/publicdomain/zero/1.0/) applies to the data made available in this article, unless otherwise stated*):

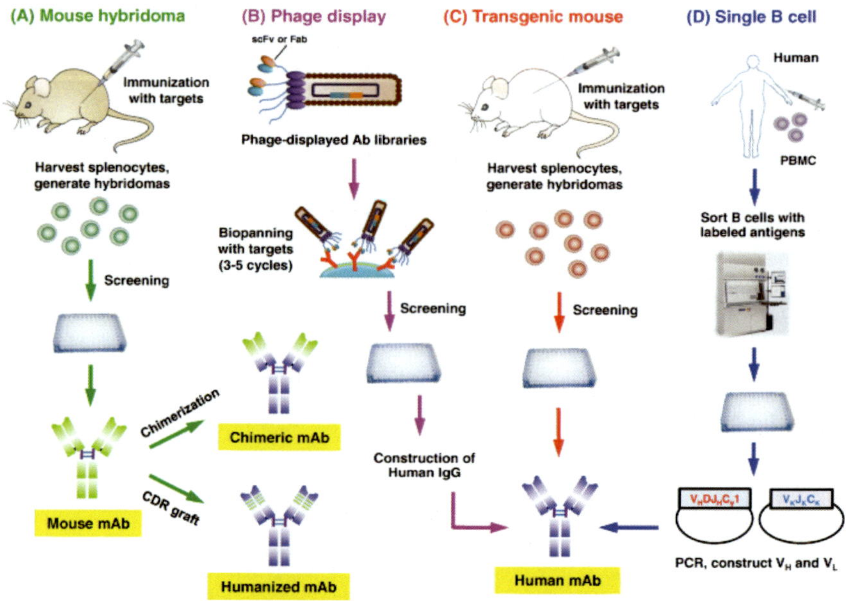

In addition to the direct targeting of maladies by mAbs via ADCC or CDC or cell death induction, there are "additions" to mAbs in the form of bispecific antibodies, antibody-drug conjugates (ADCs), chimeric antigen receptor T cell (CAR-T) therapy and antibody-radionuclide conjugates. The employment of high-throughput sequencing, robotic screening, novel biomarkers, combination with other therapeutic systems such as radiotherapy can all augment mAb-based therapy to realize the goals of biotechnology viz. optimal health for all.

"We are at the beginning of a new era of immunochemistry, namely the production of "antibody based" molecules"

— César Milstein

From Nobel Lecture (8 Dec 1984), collected in Tore Frängsmyr and Jan Lindsten (eds.), Nobel Lectures in Physiology Or Medicine: 1981-1990 (1993).

2.10 REFERENCES

1. Baah, S., Laws, M., & Rahman, K. M. (2021). Antibody-Drug Conjugates-A Tutorial Review. *Molecules (Basel, Switzerland)*, *26*(10), 2943. https://doi.org/10.3390/molecules26102943
2. Lu, RM., Hwang, YC., Liu, IJ. *et al.* Development of therapeutic antibodies for the treatment of diseases. *J Biomed Sci* 27, 1 (2020). https://doi.org/10.1186/s12929-019-0592-z
3. José F. Santos-Neto, Fabricia O. Oliveira, Katharine V. S. Hodel, Larissa M. S. Fonseca, Roberto Badaró, Bruna A. S. Machado, "Technological Advancements in Monoclonal Antibodies", The Scientific World Journal, vol. 2021, Article ID 6663708, 19 pages, 2021. https://doi.org/10.1155/2021/6663708
4. Cai HH. Therapeutic Monoclonal Antibodies Approved by FDA in 2020. Clin Res Immunol 2021;4(1):1-2.
5. Torrente-López A, Hermosilla J, Navas N, Cuadros-Rodríguez L, Cabeza J, Salmerón-García A. The Relevance of Monoclonal Antibodies in the Treatment of COVID-19. *Vaccines*. 2021; 9(6):557. https://doi.org/10.3390/vaccines9060557
6. Bryan E. Jones, Patricia L. Brown-augsburger, Kizzmekia S. Corbett, Kathryn Westendorf, Julian Davies, Thomas P. Cujec, Christopher M. Wiethoff, Jamie L. Blackbourne, Beverly A. Heinz, Denisa Foster, Richard E. Higgs, Deepa Balasubramaniam, Lingshu Wang, Yi Zhang, Eun Sung Yang, Roza Bidshahri, Lucas Kraft, Yuri Hwang, Stefanie Žentelis, Kevin R. Jepson, Rodrigo Goya, Maia A. Smith, David W. Collins, Samuel J. Hinshaw, Sean A. Tycho, Davide Pellacani, Ping Xiang, Krithika Muthuraman, Solmaz Sobhanifar, Marissa H. Piper, Franz J. Triana, Jorg Hendle, Anna Pustilnik, Andrew C. Adams, Shawn J. Berens, Ralph S. Baric, David R. Martinez, Robert W. Cross, Thomas W. Geisbert, Viktoriya Borisevich, Olubukola Abiona, Hayley M. Belli, Maren De Vries, Adil Mohamed, Meike Dittmann, Marie I. Samanovic, Mark J. Mulligan, Jory A. Goldsmith, Ching-lin Hsieh, Nicole V. Johnson, Daniel Wrapp, Jason S. Mclellan, Bryan C. Barnhart, Barney S. Graham, John R. Mascola, Carl L. Hansen, Ester Falconer. The neutralizing antibody, LY-CoV555, protects against SARS-CoV-2 infection in nonhuman primates. Science Translational Medicine, 2021: Vol. 13, Issue 593, eabf1906; DOI: 10.1126/scitranslmed.abf1906

7. Dharmage SC, Perret JL and Custovic A (2019) Epidemiology of Asthma in Children and Adults. *Front. Pediatr.* 7:246. doi: 10.3389/fped.2019.00246
8. Edris, A., De Feyter, S., Maes, T. *et al.* Monoclonal antibodies in type 2 asthma: a systematic review and network meta-analysis. *Respir Res* 20, 179 (2019). https://doi.org/10.1186/s12931-019-1138-3
9. Domingo C, Pomares X, Morón A and Sogo A (2020) Dual Monoclonal Antibody Therapy for a Severe Asthma Patient. *Front. Pharmacol.* 11:587621. doi: 10.3389/fphar.2020.587621
10. https://www.who.int/news-room/fact-sheets/detail/hiv-aids
11. Flego, M., Ascione, A., Cianfriglia, M. *et al.* Clinical development of monoclonal antibody-based drugs in HIV and HCV diseases. *BMC Med* 11, 4 (2013). https://doi.org/10.1186/1741-7015-11-4
12. Spencer DA, Shapiro MB, Haigwood NL and Hessell AJ (2021) Advancing HIV Broadly Neutralizing Antibodies: From Discovery to the Clinic. *Front. Public Health* 9:690017. doi: 10.3389/fpubh.2021.690017
13. Jaworski JP, Vendrell A and Chiavenna SM (2017) Neutralizing Monoclonal Antibodies to Fight HIV-1: On the Threshold of Success. *Front. Immunol.* 7:661. doi: 10.3389/fimmu.2016.00661
14. Walsh, S. R., & Seaman, M. S. (2021). Broadly Neutralizing Antibodies for HIV-1 Prevention. *Frontiers in immunology*, *12*, 712122. https://doi.org/10.3389/fimmu.2021.712122
15. Gaudinski MR, Houser KV, Doria-Rose NA, Chen GL, Rothwell RSS, Berkowitz N, Costner P, Holman LA, Gordon IJ, Hendel CS, Kaltovich F, Conan-Cibotti M, Gomez Lorenzo M, Carter C, Sitar S, Carlton K, Gall J, Laurencot C, Lin BC, Bailer RT, McDermott AB, Ko SY, Pegu A, Kwon YD, Kwong PD, Namboodiri AM, Pandey JP, Schwartz R, Arnold F, Hu Z, Zhang L, Huang Y, Koup RA, Capparelli EV, Graham BS, Mascola JR, Ledgerwood JE; VRC 605 study team. Safety and pharmacokinetics of broadly neutralising human monoclonal antibody VRC07-523LS in healthy adults: a phase 1 dose-escalation clinical trial. Lancet HIV. 2019 Oct;6(10):e667-e679. doi: 10.1016/S2352-3018(19)30181-X. Epub 2019 Aug 28. PMID: 31473167.
16. Mayer, K. H., Seaton, K. E., Huang, Y., Grunenberg, N., Isaacs, A., Allen, M., Ledgerwood, J. E., Frank, I., Sobieszczyk, M. E., Baden, L. R., Rodriguez, B., Van Tieu, H., Tomaras, G. D., Deal, A., Goodman, D.,

Bailer, R. T., Ferrari, G., Jensen, R., Hural, J., Graham, B. S., … and the NIAID HIV Vaccine Trials Network (2017). Safety, pharmacokinetics, and immunological activities of multiple intravenous or subcutaneous doses of an anti-HIV monoclonal antibody, VRC01, administered to HIV-uninfected adults: Results of a phase 1 randomized trial. *PLoS medicine*, *14*(11), e1002435. https://doi.org/10.1371/journal.pmed.1002435

17. Li, W., Zhu, Z., Chen, W., Feng, Y., & Dimitrov, D. S. (2017). Crystallizable Fragment Glycoengineering for Therapeutic Antibodies Development. *Frontiers in immunology*, *8*, 1554. https://doi.org/10.3389/fimmu.2017.01554
18. Zahavi, D., & Weiner, L. (2020). Monoclonal Antibodies in Cancer Therapy. *Antibodies (Basel, Switzerland)*, *9*(3), 34. https://doi.org/10.3390/antib9030034

CHAPTER 3

Antibody Drug Conjugates (ADCs): The Foundation

Contents

3.1 THE BACKGROUND:

"My first oncologist told me: 'When you read about the statistics/prognosis; remember that you are NOT a statistic, you are YOU and you are unique.' The prognosis I later read about was not good, so his statement always gave me hope. I have survived over 10 years. The prognosis was 2-3 years. He was right!!"-Shared by Karen Kockelman Schultz

According to the SEER Cancer Statistics Review, 1975-2018, National Cancer Institute. Bethesda, MD, the grimness of the situation with regard to cancer can be seen by the following incidence, death rate and survival for all races shown in table 1 (Howlader et al, 2021: *All material in this report is in the public domain and may be reproduced or copied without permission; citation as to source, however, is appreciated*):

[d] Mesotheliomas of the Pleura are included in the separate group Mesothelioma for incidence but are included in the Pleura grouping for mortality; [e] Ovary excludes borderline cases or *histologies 8442, 8451, 8462, 8472, and 8473;* [f] Due to coding changes, Brain & Nervous System mortality are no longer shown separately; [g] Rate not shown for mortality. Category did not exist in mortality coding until 1999.

Site	Incidence SEER 21 areas (2014-2018)			US Mortality from US Mortality Files, National Center for Health Statistics, Centers for Disease Control and Prevention. (2014-2018)			Survival (%) (2011-2017): Based on follow-up of patients into 2018		
	Both Sexes	Males	Females	Both Sexes	Males	Females	Both Sexes	Males	Females
All Sites	450.5	489.2	425.6	155.5	185.5	133.5	67.7	66.4	68.9
Oral Cavity & Pharynx:	11.5	17.3	6.5	2.5	3.9	1.3	66.9	66.4	68.1
Lip	0.6	0.9	0.3	0.0	0.0	0.0	90.9	91.2	89.8
Tongue	3.6	5.4	2.0	0.7	1.0	0.4	68.1	68.4	67.1
Salivary gland	1.3	1.7	1.1	0.3	0.4	0.2	72.0	66.0	80.4
Floor of mouth	0.4	0.6	0.3	0.0	0.0	0.0	51.6	49.2	57.0
Gum & other oral cavity	1.6	1.9	1.3	0.4	0.5	0.3	61.3	58.7	64.5
Nasopharynx	0.6	0.9	0.3	0.2	0.3	0.1	61.9	60.0	66.7
Tonsil	2.1	3.6	0.7	0.3	0.4	0.1	75.5	76.0	72.9
Oropharynx	0.5	0.9	0.2	0.3	0.5	0.1	49.6	51.9	40.8
Hypopharynx	0.5	0.9	0.2	0.1	0.2	0.0	37.2	36.8	38.6

Other oral cavity & pharynx	0.2	0.4	0.1	0.4	0.7	0.2	50.6	52.7	42.5
Digestive System:	80.6	98.0	66.2	40.6	51.9	31.2	43.7	41.5	46.6
Esophagus	4.2	7.3	1.7	3.9	6.9	1.4	19.9	19.3	22.3
Stomach	7.2	9.7	5.3	3.0	4.0	2.2	32.4	29.1	37.4
Small intestine	2.4	2.8	2.1	0.4	0.5	0.3	68.3	68.0	68.6
Colon & Rectum:	37.8	43.2	33.3	13.7	16.3	11.5	64.7	64.3	65.3
Colon	26.5	29.1	24.3	-	-	-	63.6	63.3	63.8
Rectum	11.3	14.1	9.0	-	-	-	67.4	66.1	69.2
Anus, anal canal & anorectum	2.0	1.6	2.3	0.3	0.2	0.3	68.7	64.0	71.4
Liver & intrahepatic bile duct	9.0	13.8	4.9	6.6	9.7	4.0	20.3	20.1	20.9
Gallbladder	1.2	0.9	1.5	0.6	0.4	0.7	19.4	20.4	18.9
Other biliary	1.8	2.3	1.5	0.5	0.5	0.4	19.3	20.2	18.2
Pancreas	13.2	15.0	11.8	11.0	12.7	9.6	10.8	10.9	10.7
Retroperitoneum	0.4	0.4	0.4	0.1	0.1	0.1	55.1	52.9	56.9
Peritoneum, omentum & mesentery	0.5	0.1	0.8	0.3	0.1	0.4	32.8	39.1	32.4
Other digestive system	0.8	0.9	0.7	0.4	0.5	0.3	7.9	6.8	9.0
Respiratory System:	56.8	66.1	49.6	39.7	49.0	32.5	24.5	22.6	26.6
Nose, nasal cavity & middle ear	0.7	0.9	0.5	0.1	0.2	0.1	58.9	58.2	60.1
Larynx	2.8	4.9	1.1	0.9	1.7	0.4	60.7	61.4	57.8
Lung & bronchus	53.1	60.1	47.9	38.5	46.9	32.0	21.7	18.1	25.4
Pleura[d]	0.0	0.0	0.0	0.1	0.1	0.0	22.0	18.5	26.5
Trachea & other respiratory organs	0.2	0.3	0.1	0.1	0.1	0.1	56.2	54.8	59.9
Bones & joints	1.0	1.2	0.9	0.5	0.6	0.4	66.8	64.5	69.8
Soft tissue (including heart)	3.5	4.2	2.9	1.3	1.5	1.2	65.0	64.7	65.2
Skin (excl. basal & squamous):	24.8	32.1	19.4	3.3	5.1	1.9	92.5	90.8	94.8
Melanoma of the skin	22.8	29.3	18.0	2.3	3.4	1.4	93.3	91.7	95.3
Other non-epithelial skin	2.0	2.9	1.4	1.0	1.8	0.5	83.7	81.0	87.6
Breast	69.1	1.3	129.1	11.1	0.3	20.1	90.2	82.4	90.3
Breast (*in situ*)	17.1	0.1	32.5	-	-	-	100.0	100.0	100.0
Female Genital System:	27.4	-	51.6	8.2	-	14.9	70.2	-	70.2

Cervix uteri	3.8	-	7.5	1.2	-	2.2	66.3	-	66.3
Corpus uteri	14.5	-	27.2	1.3	-	2.4	82.7	-	82.7
Uterus, NOS	0.5	-	0.9	1.4	-	2.5	27.7	-	27.7
Ovary[e]	5.8	-	10.9	3.7	-	6.7	49.1	-	49.1
Vagina	0.4	-	0.7	0.1	-	0.2	48.7	-	48.7
Vulva	1.4	-	2.6	0.3	-	0.6	71.1	-	71.1
Other female genital system	1.0	-	1.8	0.2	-	0.4	50.2	-	50.2
Male Genital System:	54.4	118.4	-	8.1	19.5	-	97.1	97.1	-
Prostate	50.9	111.3	-	7.8	19.0	-	97.5	97.5	-
Testis	3.0	6.0	-	0.1	0.3	-	94.9	94.9	-
Penis	0.4	0.9	-	0.1	0.2	-	64.7	64.7	-
Other male genital system	0.1	0.3	-	0.0	0.0	-	85.1	85.1	-
Urinary System:	37.0	58.1	20.3	8.2	13.1	4.5	75.8	76.5	74.3
Urinary bladder	19.7	34.2	8.5	4.3	7.4	2.1	77.1	78.5	72.6
Kidney & renal pelvis	16.4	22.6	11.2	3.6	5.3	2.3	75.6	74.9	76.8
Ureter	0.5	0.8	0.4	0.1	0.1	0.1	47.7	49.3	45.5
Other urinary system	0.4	0.6	0.2	0.1	0.2	0.1	48.8	51.6	43.2
Eye & Orbit	0.8	1.0	0.8	0.1	0.1	0.1	83.6	83.6	83.7
Brain & Nervous System:[f]	6.4	7.6	5.4	4.4	5.4	3.6	32.6	31.2	34.3
Brain	6.0	7.1	5.0	-	-	-	29.4	28.4	30.6
Cranial nerves & other nervous system	0.4	0.4	0.4	-	-	-	79.5	77.5	81.5
Endocrine System:	16.3	8.9	23.5	0.8	0.8	0.8	96.7	93.0	97.9
Thyroid	15.5	8.0	22.8	0.5	0.5	0.5	98.3	96.0	98.9
Other endocrine & thymus	0.8	0.9	0.7	0.3	0.3	0.3	65.6	66.8	64.2
Lymphoma:	22.2	26.7	18.5	5.7	7.4	4.3	75.1	74.0	76.5
Hodgkin lymphoma	2.6	2.9	2.3	0.3	0.4	0.2	88.3	87.7	89.1
Non-Hodgkin lymphoma	19.6	23.7	16.2	5.4	7.0	4.1	73.2	72.0	74.6
Myeloma	7.1	8.8	5.7	3.2	4.1	2.6	55.6	55.7	55.5
Leukemia:	14.3	18.3	11.1	6.3	8.4	4.7	65.0	65.7	64.0
Lymphocytic:	7.1	9.3	5.2	1.7	2.4	1.2	82.5	82.7	82.3

Acute lymphocytic	1.8	2.0	1.5	0.4	0.5	0.4	69.9	69.5	70.3
Chronic lymphocytic	4.9	6.7	3.5	1.1	1.6	0.8	87.2	86.8	87.8
Other lymphocytic	0.4	0.6	0.2	0.1	0.2	0.1	84.3	87.5	73.8
Myeloid & Monocytic:	6.5	8.1	5.3	3.4	4.4	2.6	43.5	43.3	43.8
Acute myeloid	4.3	5.2	3.6	2.8	3.6	2.2	29.5	28.5	30.7
Chronic myeloid	1.9	2.5	1.5	0.3	0.4	0.2	70.6	69.5	72.2
Acute monocytic	0.2	0.2	0.2	0.0	0.0	0.0	24.8	24.0	25.6
Other myeloid & monocytic	0.1	0.2	0.1	0.3	0.4	0.2	38.4	39.6	36.2
Other leukemia:	0.7	0.9	0.6	1.2	1.6	0.9	50.9	50.2	51.7
Other acute leukemia	0.2	0.3	0.2	0.4	0.5	0.3	26.9	27.9	25.7
Aleukemic, subleukemic & NOS	0.5	0.6	0.4	0.8	1.1	0.6	60.2	59.5	61.0
Kaposi Sarcoma[g]	0.5	0.9	0.1	-	-	-	73.9	73.3	80.2
Mesothelioma[g]	0.8	1.4	0.4	-	-	-	11.5	9.1	18.2
Ill-defined & unspecified	15.9	19.0	13.6	11.5	14.4	9.3	40.8	40.8	40.8

When one looks at the estimated number of cancer cases and mortalities for 2021, one can see the requirement for appropriate prevention and interventions in table 2 below (Howlader et al, 2021: *All material in this report is in the public domain and may be reproduced or copied without permission; citation as to source, however, is appreciated*):

[a] Estimated deaths for colon & rectum cancers are combined; [b] Carcinoma *in situ* of the breast accounts for about 49,290 new cases annually, and melanoma *in situ* accounts for about 101,280 new cases annually; [c] More deaths than cases suggests lack of specificity in recording underlying causes of death on death certificate.

Primary Site	Estimated New Cases			Estimated Deaths		
	Both Sexes	Males	Females	Both Sexes	Males	Females
All Sites	1,898,160	970,250	927,910	608,570	319,420	289,150
Oral Cavity and Pharynx	54,010	38,800	15,210	10,850	7,620	3,230
Tongue	17,960	13,040	4,920	2,870	1,930	940
Mouth	14,290	8,400	5,890	2,650	1,520	1,130
Pharynx	18,470	14,990	3,480	3,870	3,060	810
Other Oral Cavity	3,290	2,370	920	1,460	1,110	350
Digestive System	338,090	191,090	147,000	169,280	98,140	71,140
Esophagus	19,260	15,310	3,950	15,530	12,410	3,120
Stomach	26,560	16,160	10,400	11,180	6,740	4,440
Small Intestine	11,390	6,130	5,260	2,100	1,110	990
Colon[a]	104,270	52,590	51,680	52,980	28,520	24,460
Rectum	45,230	26,930	18,300			
Anus, Anal Canal, and Anorectum	9,090	3,020	6,070	1,430	560	870
Liver and Intrahepatic Bile Duct	42,230	29,890	12,340	30,230	20,300	9,930
Gallbladder and Other Biliary	11,980	5,730	6,250	4,310	1,770	2,540
Pancreas	60,430	31,950	28,480	48,220	25,270	22,950
Other Digestive	7,650	3,380	4,270	3,300	1,460	1,840
Respiratory System	254,170	132,910	121,260	137,040	73,340	63,700
Larynx	12,620	9,940	2,680	3,770	3,020	750
Lung and Bronchus	235,760	119,100	116,660	131,880	69,410	62,470
Other Respiratory	5,790	3,870	1,920	1,390	910	480
Bones and Joints	3,610	2,100	1,510	2,060	1,190	870
Soft Tissue	13,460	7,720	5,740	5,350	2,840	2,510
Skin (excl. basal & squamous)	115,320	68,120	47,200	11,540	7,660	3,880
Melanoma of the Skin[b]	106,110	62,260	43,850	7,180	4,600	2,580
Other non-epithelial skin	9,210	5,860	3,350	4,360	3,060	1,300
Breast[b]	284,200	2,650	281,550	44,130	530	43,600
Genital Organs	376,970	260,210	116,760	69,110	35,030	34,080
Cervix (uterus)	14,480		14,480	4,290		4,290
Endometrium (uterus)	66,570		66,570	12,940		12,940
Ovary	21,410		21,410	13,770		13,770
Vulva	6,120		6,120	1,550		1,550
Vagina and other genital organs, female	8,180		8,180	1,530		1,530
Prostate	248,530	248,530		34,130	34,130	

Testis	9,470	9,470		440	440	
Penis and other genital organs, male	2,210	2,210		460	460	
Urinary System	164,000	115,750	48,250	31,940	21,640	10,300
Urinary Bladder	83,730	64,280	19,450	17,200	12,260	4,940
Kidney and Renal Pelvis	76,080	48,780	27,300	13,780	8,790	4,990
Ureter and other urinary organs	4,190	2,690	1,500	960	590	370
Eye and Orbit	3,320	1,750	1,570	400	220	180
Brain and Other Nervous System	24,530	13,840	10,690	18,600	10,500	8,100
Endocrine System	47,200	13,730	33,470	3,290	1,620	1,670
Thyroid	44,280	12,150	32,130	2,200	1,050	1,150
Other Endocrine	2,920	1,580	1,340	1,090	570	520
Lymphoma	90,390	50,460	39,930	21,680	12,740	8,940
Hodgkin Lymphoma	8,830	4,830	4,000	960	570	390
Non-Hodgkin Lymphoma	81,560	45,630	35,930	20,720	12,170	8,550
Myeloma	34,920	19,320	15,600	12,410	6,840	5,570
Leukemia	61,090	35,530	25,560	23,660	13,900	9,760
Acute lymphocytic leukemia	5,690	3,000	2,690	1,580	900	680
Chronic lymphocytic leukemia	21,250	13,040	8,210	4,320	2,620	1,700
Acute myeloid leukemia	20,240	11,230	9,010	11,400	6,620	4,780
Chronic myeloid leukemia	9,110	5,150	3,960	1,220	680	540
Other leukemia	4,800	3,110	1,690	5,140	3,080	2,060
All Other Sites[c]	32,880	16,270	16,610	47,230	25,610	21,620

3.2 ENTER ADCS

The concept of *zauberkugel* ("magic bullet") can be ascribed to Paul Ehrlich that refers to the targeting of only a pathogen and not healthy cells. This brings in antibody–drug conjugates (ADCs) that are aimed at circumventing the challenge of traditional chemotherapy regimens viz. the targeting of cancer cells and actively dividing healthy cells attributed to the ubiquitous side-effects. Research is uncovering the potential of ADCs to amalgamate the ability of antibodies to facilitate the selective binding and the property of cytotoxic drugs to kill malignant cells (Baah S, Laws M, Rahman KM, 2021).

Figure 1 below is illustrative of ADC development across time (Baah S, Laws M, Rahman KM, 2021*: © 2021 by the authors. Licensee MDPI, Basel, Switzerland. This article is an open access article distributed under the terms and conditions of the Creative Commons Attribution (CC BY) license: https://creativecommons.org/licenses/by/4.0/*):

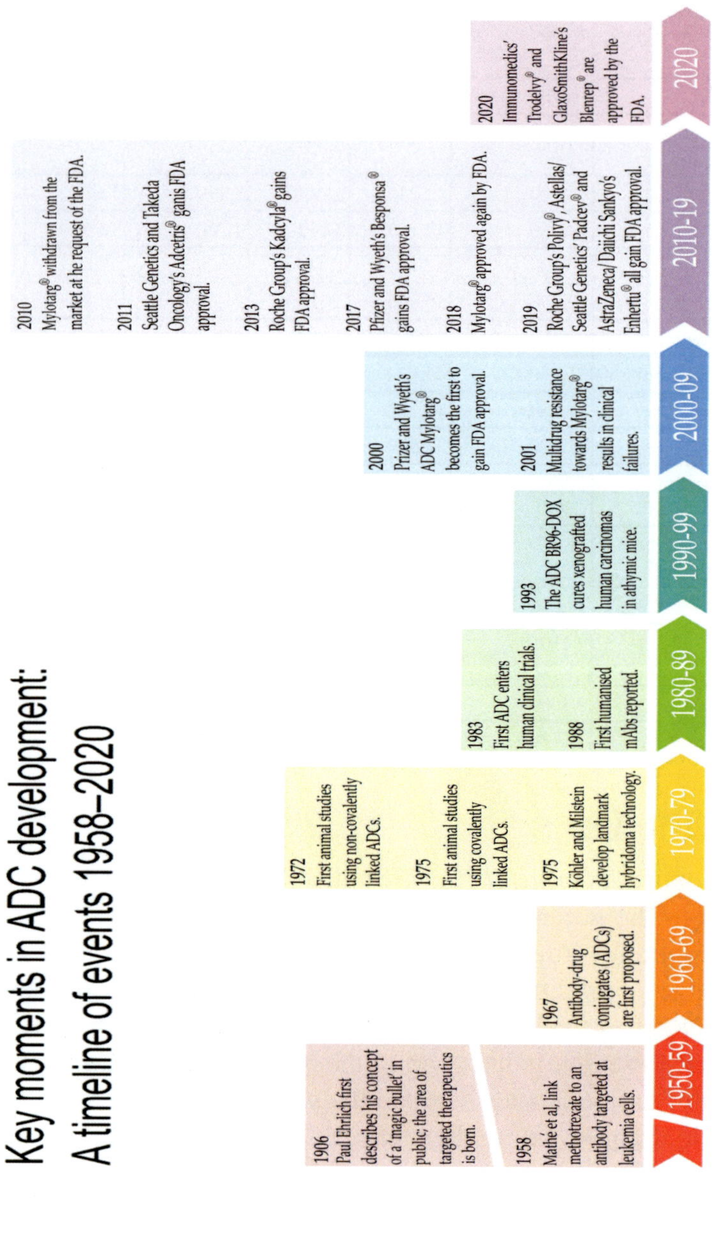

3.3 STRUCTURE OF ADCS

ADCs are composed of a monoclonal antibody, a cytotoxic drug (cytotoxic payload or warhead) and a linker that joins these two components. While mAbs are demonstrative of a slew of mechanisms ranging from immune checkpoint inhibition to complement-dependent cell-mediated cytotoxicity (CDCC) to antibody-dependent cell-mediated cytotoxicity (ADCC) to neutralization of receptor(s) and targeted signaling, ADCs entail internalization via a target receptor to deliver the cytotoxic payload inside the malignant cell (Hafeez, U., Parakh, S., Gan, H. K., & Scott, A. M, 2020; Ponziani et al, 2020; Baah S, Laws M, Rahman KM, 2021).

Figure 2 below is illustrative of an ADC (Ponziani, S., Di Vittorio, G., Pitari, G., Cimini, A. M., Ardini, M., Gentile, R., Iacobelli, S., Sala, G., Capone, E., Flavell, D. J., Ippoliti, R., & Giansanti, F (2020): © 2020 by the authors. *Licensee MDPI, Basel, Switzerland. This article is an open access article distributed under the terms and conditions of the Creative Commons Attribution (CC BY) license: http://creativecommons.org/licenses/by/4.0/*):

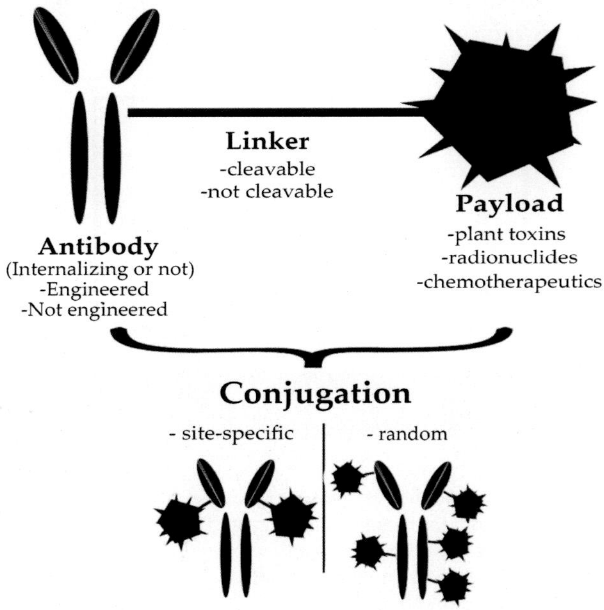

A high copy number is warranted of the ***antigen that is to be targeted*** (>10^5/cell). For instance, the ADC called ADC L49-vcMMAF targeting the p97 receptor could bind cell lines with 80,000 to 280,000 receptors per cell as opposed to other melanoma cell lines with lower expression resulting in resistance to L49-vcMMAF. Yet, other attributes like the rate of internalization and binding affinity also enter the scene. For example, the efficacy of gemtuzumab ozogamicin was more than that of trastuzumab emtansine (T-DM1) with the former binding to 5000 to 10,000 CD 33 receptors per cell and the latter to >2 million ErbB2 receptors per cell.

A vital point is the exclusive expression of this antigen to be targeted on malignant cells and comparatively lesser expression on healthy cells. Four of nine ADCs with FDA approval target cancer-specific antigens: CD22 by inotuzumab ozogamicin, CD33 by gemtuzumab ozogamicin, CD30 by brentuximab vedotin and CD79 by polatuzumab vedotin. Another aspect is the negligible secretion of a target antigen from malignant cells. This circumvents the unwanted binding of the antibody to the secreted antigens that prevents binding to the target cell (Nejadmoghaddam et al. 2019; Hafeez et al. 2020; Ponziani et al, 2020; Baah et al. 2021)

Crucial aspects of the target antigen for ADC development are summarized below in figure 3 (Nejadmoghaddam et al. 2019): © 2019 Avicenna Research Institute; *This is an open-access article distributed under the terms of the Creative Commons Attribution License, which permits unrestricted use, distribution, and reproduction in any medium, provided the original work is properly cited*):

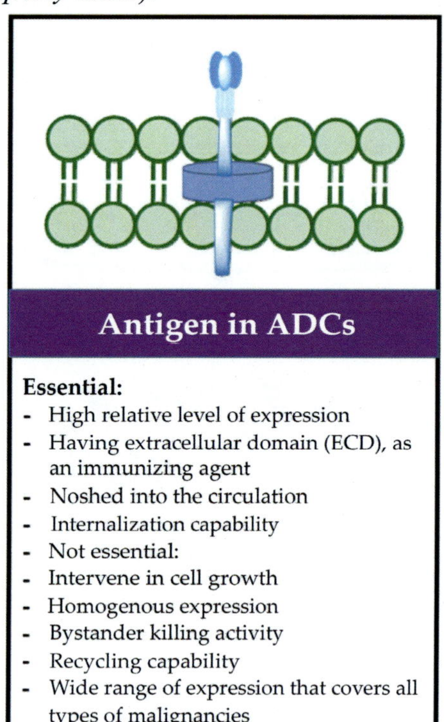

Table 3 below summarizes antigens assayed for ADCs in preclinical or clinical probing (Hafeez, U., Parakh, S., Gan, H. K., & Scott, A. M, 2020: © 2020 by the authors: Licensee MDPI, Basel, Switzerland. This article is an open access article distributed under the terms and conditions of the Creative Commons Attribution (CC BY) license: *http://creativecommons.org/licenses/by/4.0/*):

Indication	**Targets**
Acute myeloid leukaemia	CD25, CD33, CD123 (IL-3Rα), FLT3
Breast cancer	CD25, CD174, CD197 (CCR7), CD205 (Ly75), CD228 (P79, SEMF), c-MET, CRIPTO, ErbB2 (HER2), ErbB3 (HER3), FLOR1 (FRα), Globo H, GPNMB, IGF-1R, integrin β-6, PTK7 (CCK4), nectin-4 (PVRL4), ROR2, SLC39A6 (LIV1A ZIP6)
Bladder cancer	CD25, CD205(Ly75)
Colorectal cancer	CD74, CD174, CD166, CD227 (MUC-1), CD326 (Epcam), CEACAM5, CRIPTO, FAP, ED-B, ErbB3 (HER3)
Gastric cancer	CD25, CD197 (CCR7), CD228 (P79, SEMF), FLOR1(FRα), Globo H, GRP20, GCC, SLC39A6 (LIV1A ZIP6)
Gliomas GIII and GIV	CD25, EGFR
Head and neck cancer	CD71 (transferrin R), CD197 (CCR7), EGFR, SLC39A6 (LIV1A ZIP6)
Hodgkin's lymphoma	CD25, CD30, CD197 (CCR7)
Lung cancer	Axl, alpha v beta6, CD25, CD56, CD71 (transferrin R), CD228 (P79, SEMF), CD326, CRIPTO, EGFR, ErbB3 (HER3), FAP, Globo H, GD2, IGF-1R, integrin β-6, mesothelin, PTK7 (CCK4), ROR2, SLC34A2 (NaPi2b), SLC39A6 (LIV1A ZIP6)
Liver cancer	CD276 (B7-H3), c-MET
Melanoma	CD276 (B7-H3), GD2, GPNMB, ED-B, PMEL 17, endothelin B receptor
Mesothelioma	Mesothelin, CD228 (P79, SEMF)
Multiple Myeloma	CD38, CD46 (MCP), CD56, CD74, CD138, CD269 (BCMA), endothelin B receptor
Non-Hodgkin Lymphoma	CD19, CD20, CD22, CD25, CD30, CD37, CD70, CD71 (transferrin R), CD72, CD79, CD180, CD205 (Ly75), ROR1
Ovarian cancer	CA125(MUC16), CD142 (TF), CD205 (Ly75), FLOR1(FRα), Globo H, mesothelin, PTK7 (CCK4)
Pancreatic cancer	CD25, CD71 (transferrin R), CD74, CD227 (MUC1), CD228 (P79, SEMF), GRP20, GCC, IGF-1R, integrin β-6, nectin-4 (PVRL4), SLC34A2 (NaPi2b), SLC44A4, alpha v beta6, mesothelin
Prostate cancer	CD46 (MCP), PSMA, STEAP-1, SLC44A4, TENB2
Renal cancer	AGS-16, EGFR, c-MET, CAIX, CD70, FLOR1 (FRα)

The efficacy is vitally determined by an ***appropriate antibody for an ADC***; certain points to be kept in mind are inclusive of high binding affinity to the antigen, specific, low cross-reactivity, diminished immunogenicity, efficient internalization and long plasma half-life. Among the five predominant classes IgA, IgD, IgE, IgG, and IgM in humans, the IgG antibody has received the limelight in terms of therapeutic assays. While the stability and FcR binding of the IgG1 subtype has been mainstream, the concept of Fc-silent mAbs has also been reported. It can be kept in mind that the efficacy cannot be ensured by elevated antigen expression on a malignancy or that solid tumors can be effectively penetrated by ADCs with high antigen affinity. Figure 4 below summarizes these crucial aspects of the antibody for use in an ADC (Nejadmoghaddam, M. R., Minai-Tehrani, A., Ghahremanzadeh, R., Mahmoudi, M., Dinarvand, R., & Zarnani, A. H. (2019): © 2019 Avicenna Research Institute; *This is an open-access article distributed under the terms of the Creative Commons Attribution License, which permits unrestricted use, distribution, and reproduction in any medium, provided the original work is properly cited*):

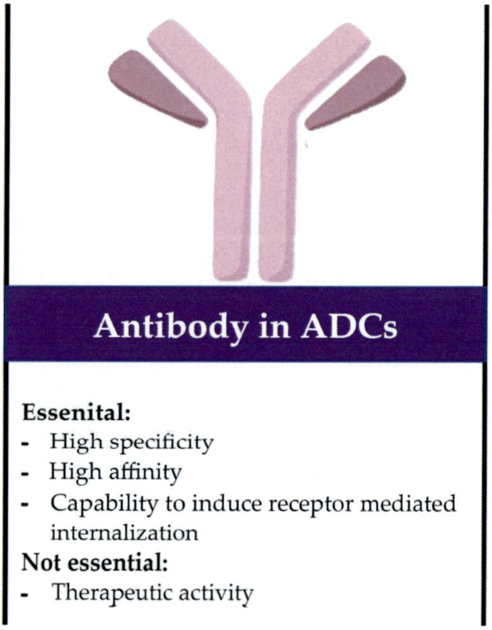

Examples of antibodies and drugs in ADCs for clinical use are summarized in table 4 below (Hafeez, U., Parakh, S., Gan, H. K., & Scott, A. M, 2020: © 2020 by the authors: Licensee MDPI, Basel, Switzerland. This article is an open access article distributed under the terms and conditions of the

Creative Commons Attribution (CC BY) license: *http://creativecommons.org/licenses/by/4.0/*):

Payload	Target Antigen	Antibody–Drug Conjugates	Antibody	Drug to Antibody Ratio	Linker	Approved Indications	Year of FDA Approval	Year of EMA Approval
Calicheamicin derivative	CD22	Inotuzumab ozogamicin (Besponsa)	Recombinant humanized IgG4	5–7	Acid-labile hydrazone-based linker	B cell precursor ALL	2017	2017
	CD33	Gemtuzumab ozogamicin (Mylotarg)	Humanized IgG4	2–3	Acid-labile hydrazone-based linker	CD33-positive AML	2000 (withdrawn 2010); reapproved 2017	2018
DM1	ErbB2	Trastuzumab emtansine (T-DM1, Kadcyla),	Humanized IgG1	3.5	Non-cleavable thioether linker	ErbB2-positive metastatic breast cancer	2013	2013
MMAE	CD30	Brentuximab vedotin (SGN-35, Adcetris)	Chimeric IgG1	4	Protease-cleavable linker	Hodgkin's lymphoma, ALCL, PTCL, mycosis fungoides	2011	2012
	CD79	Polatuzumab vedotin (Polivy)	Humanized IgG1	4	Protease-cleavable linker	DLBCL	2019	Not approved by the EMA
	Nectin-4	Enfortumab vedotin (ASG-22ME, Padcev)	Human IgG1	4	Cleavable valine-citrulline linker	Advanced urothelial cancer	2019	Not approved by the EMA
MMAF	BCMA	Belantamab mafodotin (GSK2857916, Blenrep)	Humanized IgG1	Unknown	Non-cleavable maleimidocaproyl (mc) linker	Relapsed/refractory multiple myeloma	2020	Orphan drug designation by the EMA, 2017
DXd (DX-8951 derivative)	ErbB2	Trastuzumab deruxtecan (DS-8201a, Enhertu)	Humanized IgG1	8	Cleavable peptide linker	Metastatic ErbB2-positive breast cancer	2019	Not approved by the EMA
SN-38	TROP2	Sacituzumab govitecan (IMMU-132, Trodelvy)	Humanized IgG1	7.6	Cleavable CL2A linker	Triple-negative breast cancer	2020	Not approved by the EMA

Let us now look at the cytotoxic payloads in an ADC. These should have:

a) an appropriate functional group for conjugation to the antibody
b) highly super-toxic to malignant cells at minimal payload delivery
c) optimal solubility for conjugation and ADC functioning under physiological conditions
d) elevated cytotoxic activity (half maximal inhibitory concentration (IC_{50}) = 0.01 to 0.1 nM) over standard chemotherapeutic molecules
e) stability during storage and transport in the body to circumvent the formation of unwanted products due to instability

Figure 5 below summarizes key requisites of payloads in an ADC (Nejadmoghaddam, M. R., Minai-Tehrani, A., Ghahremanzadeh, R., Mahmoudi, M., Dinarvand, R., & Zarnani, A. H. (2019): © 2019 Avicenna Research Institute; *This is an open-access article distributed under the terms of the Creative Commons Attribution License, which permits unrestricted use, distribution, and reproduction in any medium, provided the original work is properly cited*):

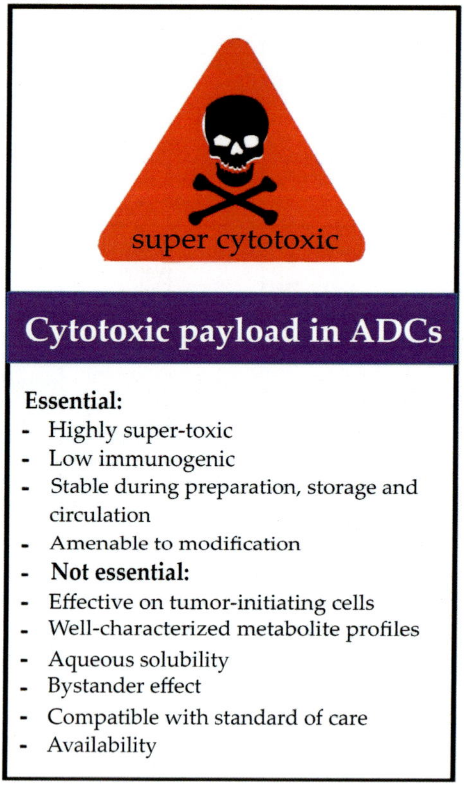

There are two predominant toxic compound families: that target microtubules (auristatins and maytansines) and those that target the DNA (calicheamicins and pyrrolobenzodiazepines or PBDs). A third class encompasses drugs targeting specific enzymes for cell survival (such as amatoxins targeting RNA polymerase II and thailanstatins targeting the spliceosome). Given that malignancies are exemplified by rapid division, microtubule-targeting molecules find predominant use in ADCs. These can be tubulin polymerization inhibitors (microtubule destabilizers) or tubulin polymerization promoters (microtubule stabilizers). Figure 6 below illustrates ***TOP***: the functioning and ***BOTTOM***: structures of tubulin-targeting payloads (Ponziani, S., Di Vittorio, G., Pitari, G., Cimini, A. M., Ardini, M., Gentile, R., Iacobelli, S., Sala, G., Capone, E., Flavell, D. J., Ippoliti, R., & Giansanti, F (2020): © 2020 by the authors. Licensee MDPI, Basel, Switzerland. This article is an open access article distributed under the terms and conditions of the Creative Commons Attribution (CC BY) license: *http://creativecommons.org/licenses/by/4.0/*):

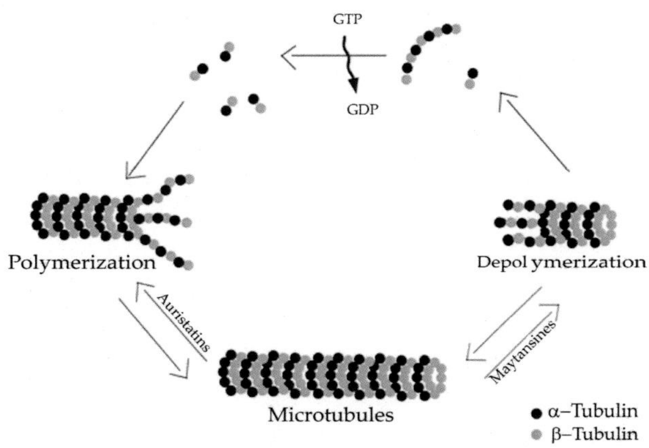

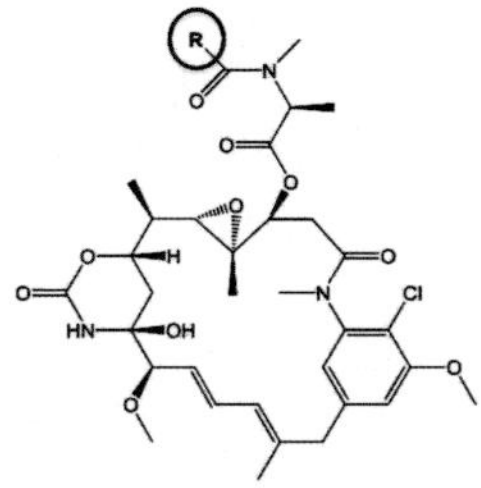

Chemical structures of maytansinoid analog payloads.

- **N(2')-deacetyl-N(2')-(3-mercapto-1-oxopropyl)-maytansine (DM1).** **DM1: R=CH_2-CH_2-SH**
- **N(2')-deacetyl-(N2')-(4mercapto-1-oxopentyl)-maytansine (DM3).** **DM3: R=CH_2-CH_2-$CH(CH_3)$-SH**
- **N(2')-deacetyl-N(2')-(4-mercapto-4-methyl-1-oxopentyl)-maytansine (DM4).** **DM4: R=CH_2-CH_2-$C(CH_3)_2$-SH**

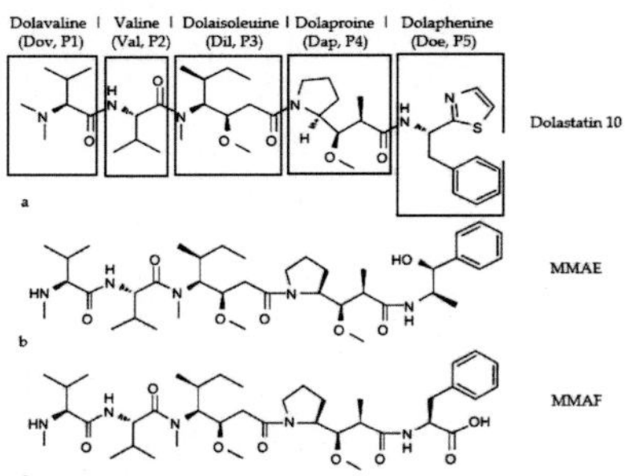

Chemical structures of:
dolastatin 10 (a) and auristatin analogues:
monomethyl auristatin E (b),
monomethyl auristatin F (c)

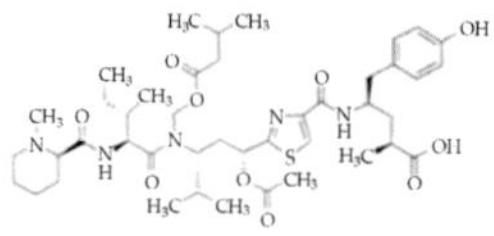

Tubulysin A

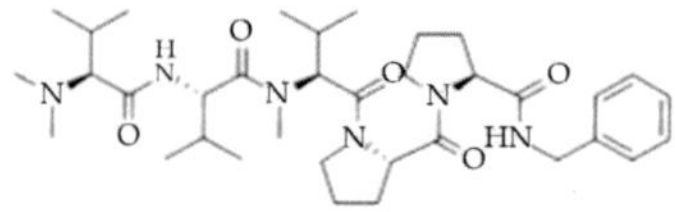

Cemadotin

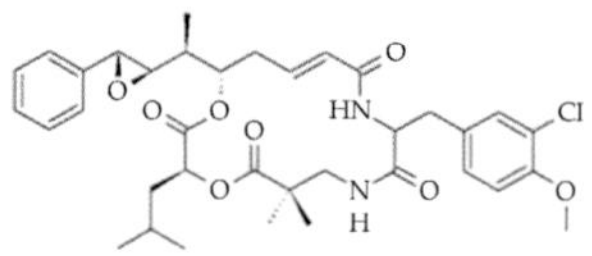

Cryptophycins

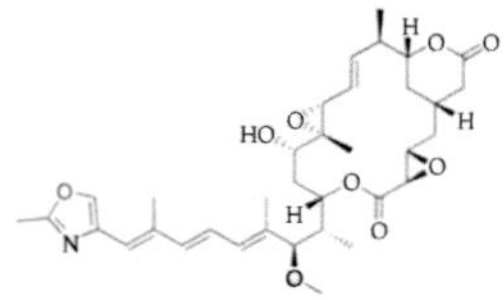

Rhizoxin

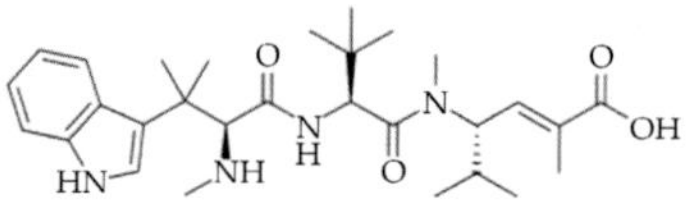

Hemiasterlin

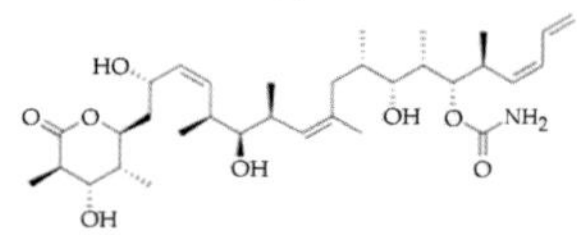

Discodermolide

Looking at molecules that target DNA, two possible advantages emerge over the previous class; the first is the ability to target non-dividing cells (CSCs-discussed below) and picomolar IC_{50} values over nano-scale IC_{50} values for microtubule-targeting drugs. These molecules cause double-strand breakage, cross-linking, alkylation and intercalation as depicted below in figure 7 (Ponziani, S., Di Vittorio, G., Pitari, G., Cimini, A. M., Ardini, M., Gentile, R., Iacobelli, S., Sala, G., Capone, E., Flavell, D. J., Ippoliti, R., & Giansanti, F (2020): © 2020 by the authors. Licensee MDPI, Basel, Switzerland. This article is an open access article distributed under the terms and conditions of the Creative Commons Attribution (CC BY) license: *http://creativecommons.org/licenses/by/4.0/*):

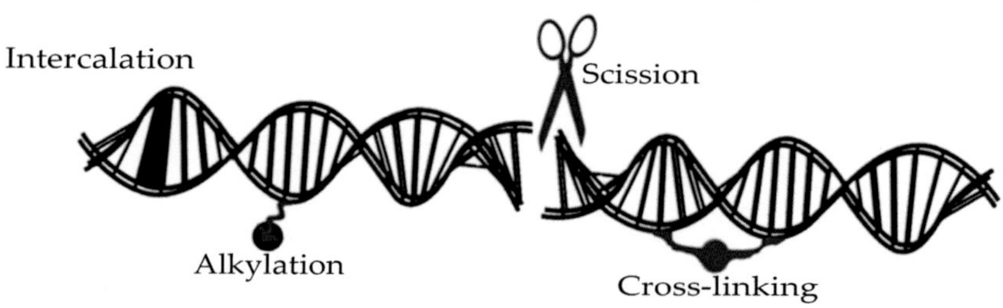

These molecules are inclusive of *Streptomyces* sp-derived PBDs and duocarmycins, calicheamicins and camptothecin (CPT) from *Camptotheca acuminata* along with the analog DUBA (duocarmycin-hydroxybenzamide-azaindole) as shown below in figure 8 (Ponziani et al. 2020): © 2020 by the authors. Licensee MDPI, Basel, Switzerland. This article is an open access article distributed under the terms and conditions *of the Creative Commons Attribution (CC BY) license: http://creativecommons.org/licenses/by/4.0/*):

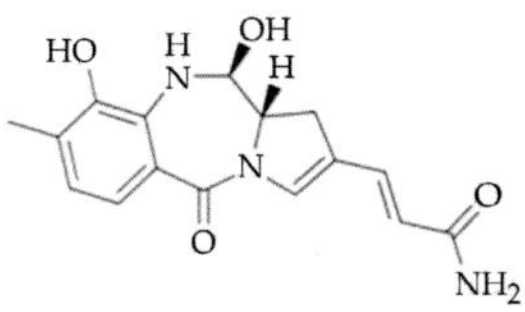

Pyrrolobenzodiazepines

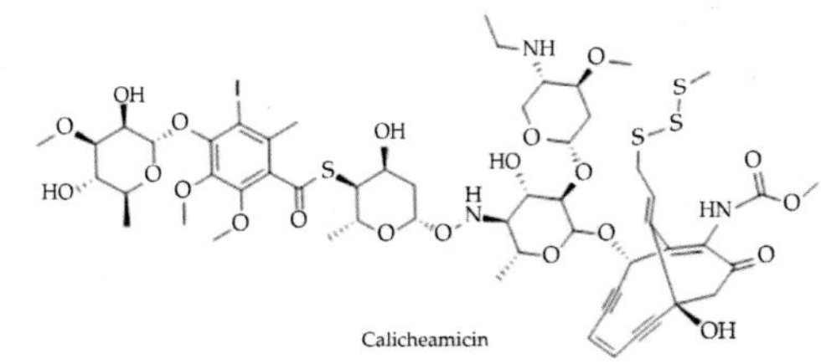

Calicheamicin

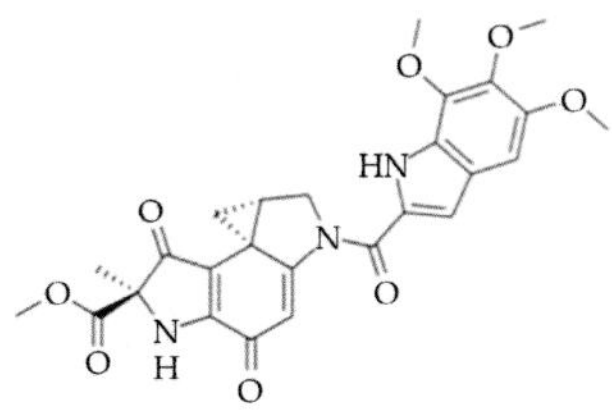

Duocarmycin A

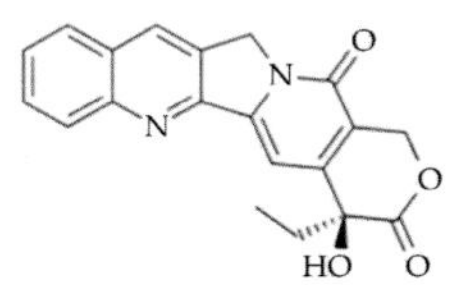

Camptothecin

Figure 9 below summarizes all the approaches employed so far in ADCs (Ponziani, S., Di Vittorio, G., Pitari, G., Cimini, A. M., Ardini, M., Gentile, R., Iacobelli, S., Sala, G., Capone, E., Flavell, D. J., Ippoliti, R., & Giansanti, F (2020):

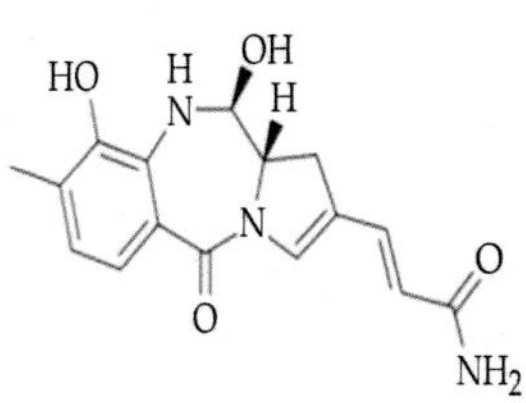

Pyrrolobenzodiazepines

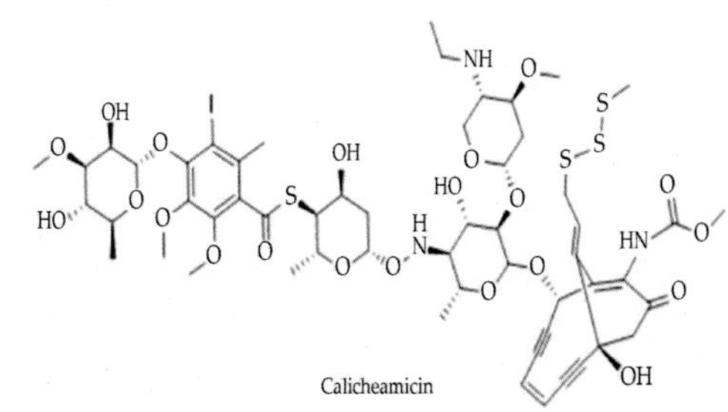

Calicheamicin

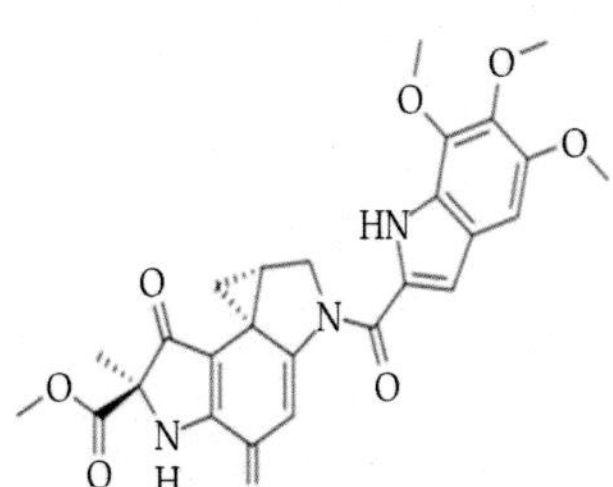

Duocarmycin A

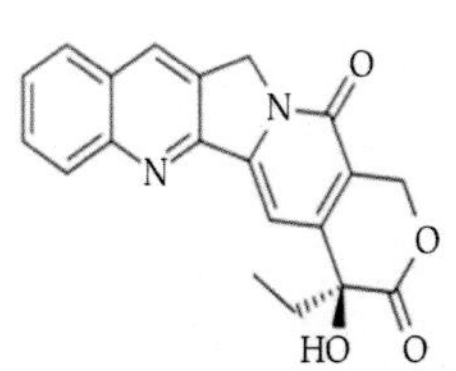

Camptothecin

Now, we shall look at the ***linkers***: this becomes dynamic research as the correct design impacts the efficacy. This is inclusive of the stability of the linker employed to circumvent unintended payload release. Cleavable linkers are when the payload is released at the tumor site whereas in non-cleavable linkers, there is no separation of the mAb and the payload-the former is degraded post-internalization.

It is vital for cleavable linkers not to show bystander killing. Such selective release entails a shift in pH across compartments or the presence of certain enzymes in the cancer cell cytoplasm. There are disulfide linkers that are sensitive to elevated glutathione levels in malignant cells and are reduced by the latter to facilitate the release of the payload. Commercial ADCs like Mylotarg® and Besponsa® contain hydrazone linkers that are stable in the bloodstream at pH 7.4 and undergo hydrolysis in an acidic cellular compartment viz. lysosome (pH < 5) or late endosomes (pH 5.5–6.2). There are dipeptide linkers (enzyme cleavable linkers) that are targeted by cathepsin B (a lysosomal protease) present in elevated levels in tumor cells. Examples of dipeptides are alanine–alanine in ImmunoGen's IMGN632, valine–alanine in loncastuximab tesirine of ADC Therapeutics and valine–citrulline in Adcetris®. These dipeptides and the payload are separated by a *para*-aminobenzyloxycarbonyl (PABC) spacer unit.

The structure of p-aminobenzyl carbamate, CAS#:918132-66-8 (PABC) is depicted in figure 10 below (Ponziani et al. 2020):

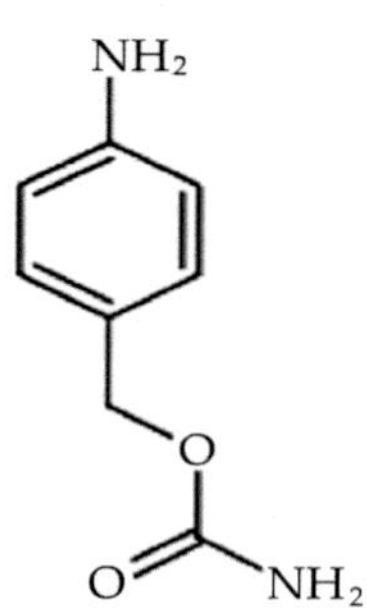

The types of cleavable linkers for ADCs are illustrated in figure 11 below (Ponziani et al. 2020):

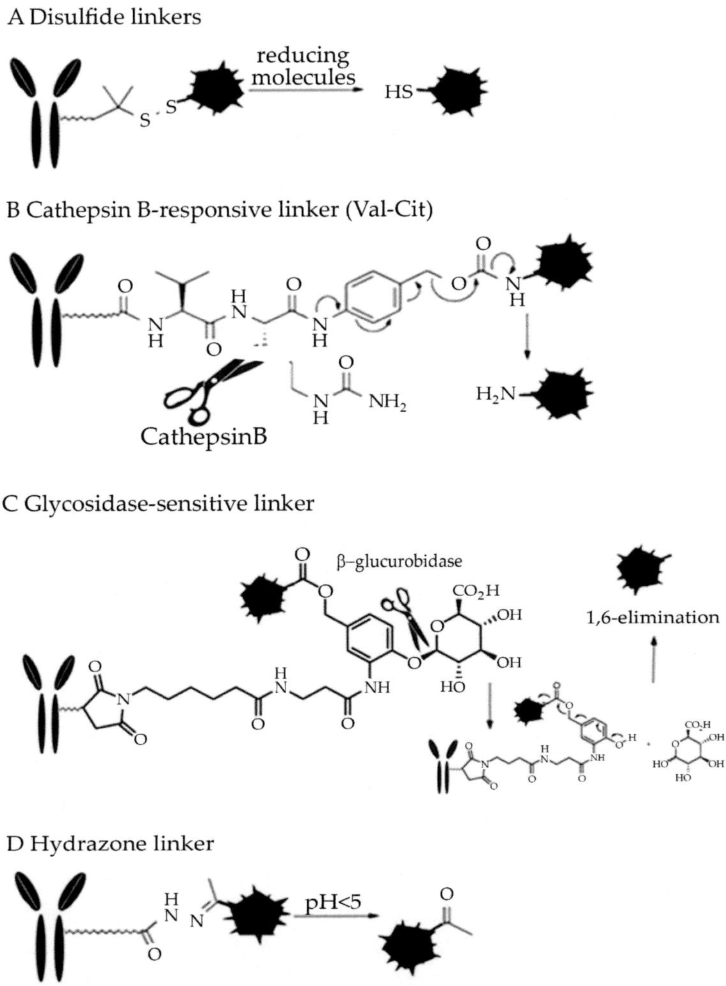

Non-cleavable linkers are alkylic and polymeric with a thioether or maleimidocaproyl group. They do not undergo fragmentation and release the payload after the mAb has been digested by lysosomes pos-antigen internalization. This facilitates lower off-target toxicity (bystander killing) due to augmented stability in the blood. Examples are maleimidocaproyl (MC) in depatuxizumab mafodotin from AbbVie and monomethyl auristatin E (MMAE) or monomethyl auristatin F (MMAF)-containing payload ADCs.

Illustrative linkers are depicted in figure 12 below: (a) A cleavable hydrazone linker and disulphide trigger of Mylotarg®; (b) an MC–Val–Cit–PABC linker with the non-cleavable MC and dipeptide cleavable Val–Cit linkers and a PABC spacer in Adcetris®; (c) a non-cleavable MCC linker of Kadcyla® (Baah S, Laws M, Rahman KM, 2021*: © 2021 by the authors. Licensee MDPI, Basel, Switzerland. This article is an open access article distributed under the terms and conditions of the Creative Commons Attribution (CC BY) license: https://creativecommons.org/licenses/by/4.0/*):

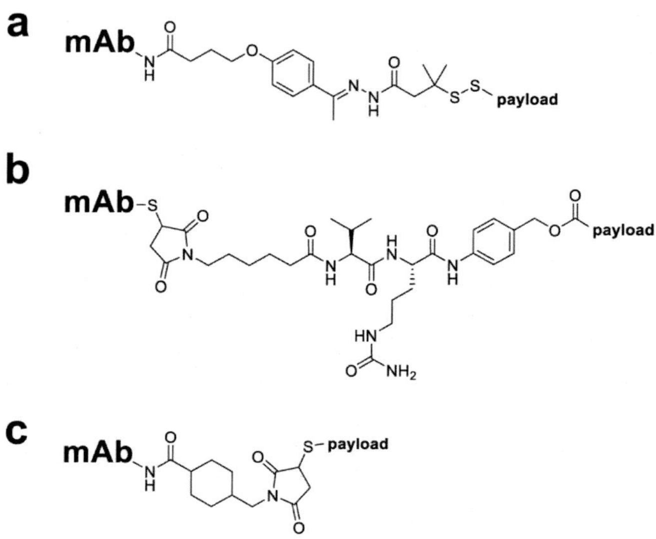

The various aspects of linker design for ADCs are summarized in figure 13 below (Nejadmoghaddam, M. R., Minai-Tehrani, A., Ghahremanzadeh, R., Mahmoudi, M., Dinarvand, R., & Zarnani, A. H. (2019): © 2019 Avicenna Research Institute; *This is an open-access article distributed under the terms of the Creative Commons Attribution License, which permits unrestricted use, distribution, and reproduction in any medium, provided the original work is properly cited*):

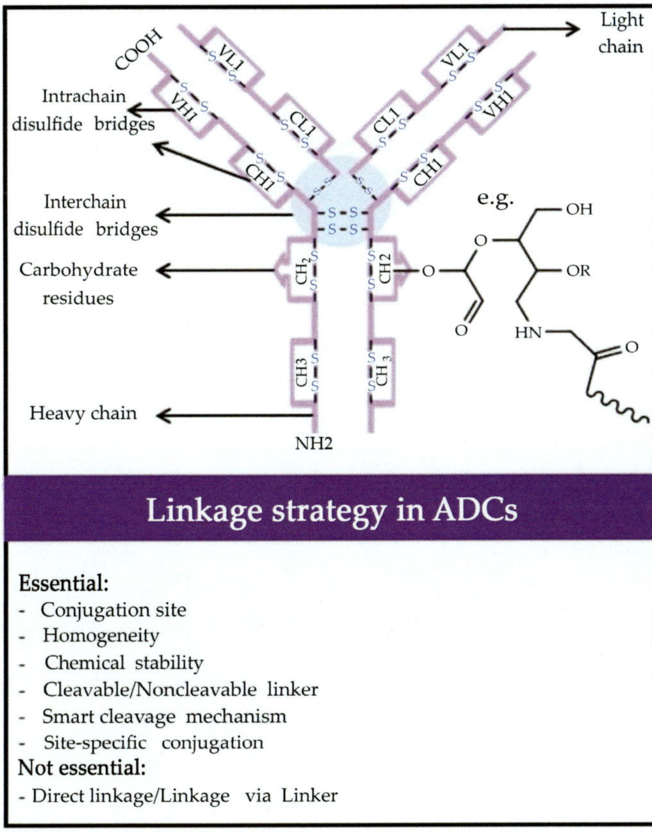

3.4 ADC UPTAKE

Receptor-mediated endocytosis (RME) is employed by internalizing ADCs. This can be followed by the receptor-mediated endocytosis via clathrin-mediated and caveolae-mediated endocytosis (CME) or can be independent of the receptor or clathrin-caveolin-independent endocytosis. This is followed by the formation of early endosomes to form late endosomes for lysosomal fusion and payload release.

Figure 14 below displays the scheme of **a** - ADC structure **b** - ADC trafficking and processing **c**- endocytosis and autophagy (Chalouni, C., Doll, 2018: *Open Access: This article is distributed under the terms of the Creative Commons Attribution 4.0 International License (http://creativecommons.org/licenses/by/4.0/), which permits unrestricted use, distribution, and reproduction in any medium, provided you give appropriate credit to the original author(s) and the source, provide a link to the Creative Commons* *license, and indicate if changes were made. The Creative Commons Public Domain Dedication waiver (http://creativecommons.org/publicdomain/zero/1.0/) applies to the data made available in this article, unless otherwise stated*):

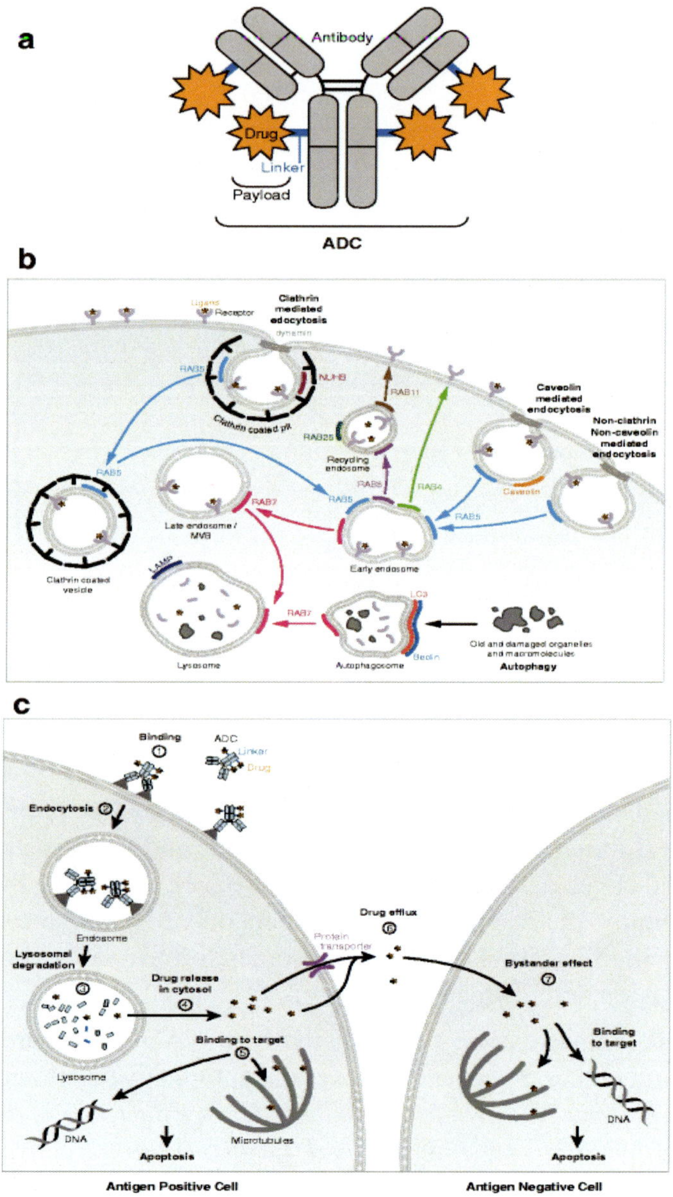

In the case of non-internalizing antibodies, the arrival of an ADC at the site of a malignancy results in altered conditions or enzyme action in the extracellular environment for payload release. This drug is then taken up by pinocytosis or diffusion. A challenge here is the bystander killing when the drug diffuses to adjacent healthy cells.

The scheme of how ADCs act is depicted in figure 15 below (Ponziani, S., Di Vittorio, G., Pitari, G., Cimini, A. M., Ardini, M., Gentile, R., Iacobelli, S., Sala, G., Capone, E., Flavell, D. J., Ippoliti, R., & Giansanti, F (2020):

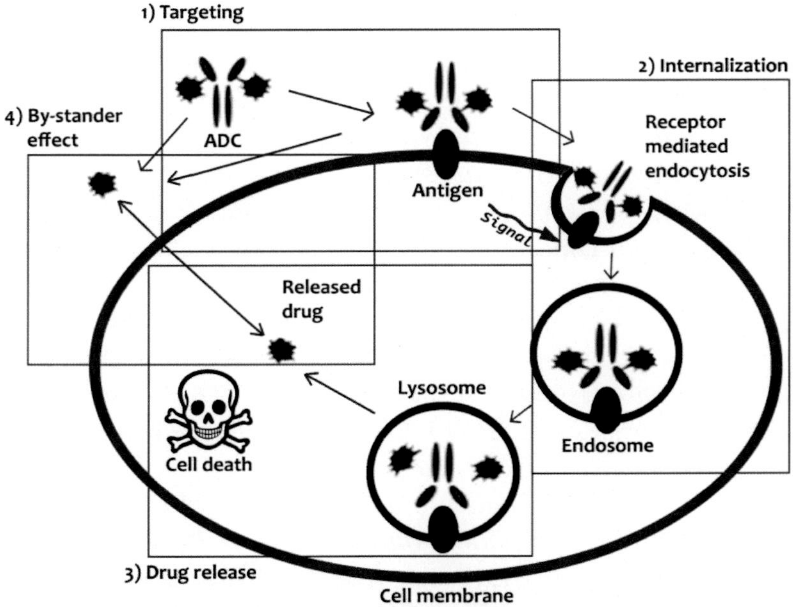

(Nejadmoghaddam et al. 2019; Hafeez et al. 2020; Ponziani et al, 2020; Baah et al. 2021)

3.5 THE CLINICAL SCENARIO

Table 5 below is an exhaustive summary of ADCs in clinical development as published in the Avicenna Journal of Medical Biotechnology (AJMB) (Nejadmoghaddam et al.2019):

Key: Not available (n/a), Relapsed B-cell non-Hodgkin's lymphoma (B-NHL), Acute myeloid leukemia (AML), Mertansine (DM1), Calicheamicin (calich.), N-succinimidyl 4-(N-maleimidomethyl) cyclohexane-1carboxylate (SMCC), Hydrazone acetyl butyrate (AcBut), Uterine Serous Carcinoma (USC), Tumor-Associated Antigen (TAA), Valine-citrulline-seco (vcseco), Renal Cell Carcinoma (RCC), clear cell Renal Cell Carcinoma (ccRCC), Mantle-Cell Lymphoma Diffuse (MCLD), Non Small-Cell Lung Cancer (NSCLC), Receptor tyrosine kinases (RTKs), Recurrent Glioblastoma Multiforme (GBM), Transmembrane Protein (TP), CD27 ligand (CD27L), Epidermal growth factor receptor variant III (EGFRvIII), Glioblastoma multiforme (GBM), Epithelial Ovarian Cancer (EOC), Head and Neck Squamous Cell Carcinomas (HNSCC), Auristatin F-hydroxypropylamide (AF-HPA), Polyacetal-based polymer (Fleximer®), Non-natural amino acid linker para-acetyl-phenylalanine (*pAcF), Amberstatin, a short polyethylene glycol (PEG) spacer terminated by an alkoxyamine (AS269); B Cell Receptor (BCR), Chronic Lymphocytic Leukemia (CLL), Prostate-specific membrane antigen (PSMA), Maleimido-[short PEG]-Lys-PABOCO-20-O (CL2A), Metastatic colorectal cancer (mCRC), Carcinoembryonic Antigen Related Cell Adhesion Molecule 5 (CEACAM5), Trophoblast cell-surface antigen 2 (Trop-2), Tumor-Associated Calcium Signal Transducer (TACSTD2), Gastric Antigen 733-1 (GA733-1), Malignant Pleural Mesothelioma (MPM), Platinum-resistant ovarian cancer (P-OC); Target sodium phosphate transporter 2b (NaPi2b), Transmembrane cell surface receptor guanylyl cyclase C (GCC), Delta-like protein 3 (DLL3), polyethylene glycol spacer (PEG8), Selective Catalytic Reduction (scr), Metastatic Urothelial Cancer (MUC), B-Cell Maturation Antigen (BCMA), DX-8951 a derivative of the camptothecin analog exatecan (DXd); Folate receptor 1(FOLR1), Maleimidocaproyl-valine-citrulline-(MC-VC-PABC), Carbohydrate antigen 125 (CA-125), Mucin 16 (MUC16), A high molecular weight mucin-type glycoprotein (CanAg), Erythropoietin producing hepatoma A2 receptor (EphA2 or EPHA2), Ectonucleotide pyrophosphatase/phosphodiesterase family member 3 (ENPP3), Poliovirus receptor related protein 4 (PVRL4), 2 N-terminal Leucine-Rich Repeat (LRR), Human Tissue Factor (TF), Stem Cell Factor Receptor c-Kit (SCFR); Hepatocyte Growth Factor Receptor (HGFR), Structural homolog of the urokinase-type Plasminogen Activator Receptor (uPAR), Tumor-associated antigen (C4.4a), Lung Squamous Cell Carcinoma (LSCC), Fibroblast growth factor receptor type 2 (FGFR2), Ovarian Cancers (OC), Trophoblast Glycoprotein (TPBG), metastatic Castration-Resistant Prostate Cancer (mCRPC), - transmembrane epithelial antigen of the prostate-1 (STEAP1), Anti-solute carrier family 39 zinc transporter member 6 (SLC39A6; LIV-1; ZIP6), Anti-Endothelin B Receptor (ETBR), Auristatin-0101 (Aur0101); Lymphocyte antigen 6 complex locus E (Ly6E), Antigen-Presenting Cell (APC), α subunit of the interleukin-2 receptor (IL-2R alpha), Lysosome-Associated Membrane Protein 1 (LAMP1)*

ADC names	**Clinical phase, indication**	**Ab, kd, therapeutics activity**	**Payload**	**Linkage strategy**	**DAR, MTD, bystander effect**	**Sponsor, Reference**
Targeting HER2 antigen, a transmembrane RTKs in the growth of some cancer cells:						
Kadcyla, Ado-Trastuzumab emtansine, T-DM1	Approved in 2013, for treatment of her2 positive breast cancer	huIgG1 (trastuzumab), n/a, ADCC and HER2-dependent PI3K/AKT signaling	DM1	Native lysine residues, SMCC nonreducible thioether linkage	~3.5, 3.6 *mg/kg*, no	Genentech, Inc. (8–11)
SYD985, Trastuzumab vc-seco-DUBA	Phase I, for treatment of USC and epithelial EOC	huIgG2 anti HER2 (Trastuzumab), n/a, no	DUO	VC-seco	~ 2.8, 1.88 *mg/kg*, yes	Synthon BV (37–39)
ADC XMT-1522	Phase I, for treatment of low HER+ breast, gastric and lung cancers	huIgG1anti-HER2 (HT-19), n/a, n/a	AF-HPA	Fleximer®	12, n/a, yes	Mersana Therapeutics (40)
ADC ARX788	Phase I, for treatment of low HER+ breast, ovarian, lung and gastric cancers	IgG1anti-HER2, n/a, n/a	MMAF	pAcF site-specific oxime linkage, AS269 noncleavable linker	2, n/a, n/a	Zhejiang Medicine/Ambrx (41)
ADC ADCT-502	Phase I, for treatment of low HER2+ expressing breast, NSCLC, gastro-esophageal, bladder cancer	huIgG1anti-HER2 (trastuzumab)	PBD	Cysteine residues, VA-PABC	1.7, n/a, n/a	ADC Therapeutics S.A. (42)
Targeting EGFR antigen, a RTKs that is essential for ductal and lobuloalveolar development:						
ABT-414, Depatuxizumab mafodotin	Phase II, for treatment of GBM	huIgG1 anti EGFR (ABT-806), 0.06 *nM*, inhibits EGFR signaling	MMAF	Native cysteine residues, MC noncleavable linker	~3.8,1.5 *mg/kg*, no	Abbvie (43)
AMG 595	Phase I, for treatment of GBM	huIgG1anti-EGFRvIII, 0.61 *nM*, n/a	DM1	Native lysine residues, SMCC noncleavable thioether linker	~3.5, n/a, no	Amgen (44)
IMGN289, Laprituximab emtansine	Phase I, for treatment of NSCLC and HNSCC	huIgG anti-EGFR (J2898A), n/a, n/a	DM1	Native lysine residues, SMCC noncleavable thioether linker	n/a, n/a, no	ImmunoGen (45)
ABBV-221	Phase I, for treatment of solid tumor	huIgG1 anti-EGFR, n/a, n/a	MMAE	VC protease-cleavable linker	n/a, n/a, n/a	Abbvie (46)
Targeting CD70 (CD27L) antigen a TP2 and member of the tumor necrosis factor family:						

SGN-75	Phase I, for treatment of CD70-positive NHL and metastatic RCC	hu anti-CD70 (h1F6), n/a, n/a	MMAF	Native cysteine residues, MC noncleavable linker	n/a, 3, n/a	Seattle Genetics (47)
MDX-1203, BMS-936561	Phase I, for treatment of ccRCC or B-NHL	hu anti-CD70, n/a, n/a	DUO	Native cysteine residues, VC protease-cleavable linker	n/a,15 *mg/kg*, yes	Bristol-Myers (48)
SGN-CD70A	Phase I, for treatment of RCC, MCLD, LBC, FL,	hu anti-CD70, n/a, n/a	PBD	VA linker	n/a, n/a, yes	Seattle Genetics (49)
AMG 172	Phase I, for treatment of ccRCC	huIgG1, n/a, n/a	DM1	Native lysine residues, MCC noncleavable linker	n/a, n/a, no	Amgen (50)
Targeting CD33 antigen, a EGP:						
Mylotarg, Gemtuzumab Ozogamicin (GO)	Withdrawn 2010 and approved in 2017, for treatment of CD33⁺ AML	huIgG4, n/a, n/a	Calich.	Native lysine residues, (AcBut)-N-acyl acid-labile hydrazone linker	n/a, 0.25 *mg/kg*, yes	Pifizer (51)
SGN-CD33A	Phase I, for treatment of AML	hu anti-CD33 with engineered cysteines, n/a, n/a	PBD	Engineered cysteine residues, VA linker	n/a, n/a, yes	Seattle Genetics (12,13)
AVE9633	Phase I, for treatment of AML	anti-CD33, n/a, n/a	DM4	Native lysine residues, SPDB disulfide cleavable linker	n/a, n/a, n/a	Sanofi (53)
Targeting CD19 antigen, a TP1 on B cells as an accessory molecule for B-cell signal transduction and TAA:						
SAR3419, coltuximab ravtansine	Phase II, for treatment of B-NHL and B-ALL	huIgG1 anti-CD19 (huB4), n/a, ADCC	DM4	Native lysine residues, SPDB disulfide cleavable linker	~3.5, ~4.3 *mg/kg*, yes	ImmunoGen (7,34,35)
SGN-CD19A	Phase I, for treatment of B-Cell Malignancies	huIgG1 anti-CD19 (hBU12), n/a, ADCC	MMAF	Native cysteine residues, MC linker, noncleavable	n/a, 6.0, no	Seattle Genetics (32)
ADCT-402	Phase I, for treatment of relapsed or refractory B-ALL	huIgG1anti-CD19, n/a, n/a	PBD	Native cysteine residues, VA and maleimide cleavable linker	n/a, n/a, n/a	ADC Therapeutics S.A. (33)
Targeting Mesothelin antigen, a glycophosphatidyl inositol anchored protein:						
BAY 94–9343, anetumab ravtansine	Phase II, for treatment of MPM	hu anti-mesothelin, n/a, n/a	DM4	Lysine residues, SPDB disulfide cleavable linker	n/a, 6.5 *mg/kg*, yes	Bayer (57)
BMS-986148	Phase I & II, for treatment of Mesothelin-expressing cancers	Anti-mesothelin	n/a	n/a	n/a, n/a, n/a	Bristol-Myers (58)
DMOT4039A	Phase I, for treatment of pancreatic and P-OC	hu anti-mesothelin (7D9.v3), n/a, n/a	MMAE	A noncleavable alkyl hydrazide linker	~ 3.5, 2.4 *mg/kg*, n/a	Genentech, Inc. (59,60)
Targeting CD22 antigen, a transmembrane sialoglycoprotein functions as an inhibitory receptor for BCR signaling and BCR-induced cell death:						

Inotuzumab, IO, Ozogamicin, CMC-544	Approved in 2017, for treatment of CD22⁺ ALL	huIgG4 anti CD29(G544), n/a, no	Calich.	Native lysine residues, (AcBut)-N-acyl, Acid-labile hydrazone linker	n/a, 0.05 *mg/kg*, yes	Pfizer (12)
Pinatuzumab vedotin, DCDT2980S, RG7593	Phase II, for treatment of NHL and CLL	huIgG1anti-CD22 (Epratuzumab), n/a, n/a	MMAE	Native cysteines residues, MC-VC-PAB linker	~ 2.4, 2.4 *mg/kg*, yes	Genentech, Inc. (61)
Targeting CEACAM5 antigen, labetuzumab, CEA, CD66e, a EGP that has a role in cell adhesion and invasion:						
IMMU-130, hMN14-SN38, labetuzumab govitecan, labetuzumab-SN-38	Phase II, for treatment of mCRC	huIgG1 anti-CEACAM5 (hMN14), 1.5 *nM*, ADCC	SN-38	Native cysteine residues, CL2A pH sensitive (Benzylcarbonate site) carbonate linker	7-8, 6–10 *mg/kg*, yes	Immunomedics (63–65)
SAR40870	Phase I & II, for treatment of B-Cell Malignancies	huIgG1 anti-CEACAM5, n/a, n/a	DM4	Lysine residues, SPDB disulfide cleavable linker	n/a, n/a, yes	Sanofi (66)
Targeting Trop-2 (M1S1, TACSTD2 or GA733-1) antigen, a EGP transduces calcium signal has a role in ERK1/2 MAPK pathway which mediates cancer cell proliferation, migration, invasion, and survival:						
IMMU-132, hrS7-SN-38, Sacituzumab govitecan	Phase III, for treatment of pancreatic cancers, SCLC and TNBC	huIgG1 anti-trop-2 (RS7 or Sacituzumab), 0.564 *nM*, ADCC	SN-38	Native cysteine residues, CL2A pH sensitive carbonate link	~7.6, 8–10 *mg/kg*, yes	Immunomedics (67–72)
PF-06664178, Trop-2 ADC, RN927C	Phase I, for treatment of OC, NSCLC and breast cancer	Engineered huIgG1anti-Trop-2, 14 *nM*, n/a	PF06380101	Site-specific transglutaminase tag, AcLys-VC-PABC linker	2.0, n/a, n/a	Pfizer (73)
Targeting PSMA antigen, a TP2 has known enzymatic activities and acts as a glutamate-preferring carboxypeptidase:						
PSMA ADC	Phase I & II, for treatment of prostate cancer	hu anti-PSMA, 35.6–46.5 *nM*, n/a	MMAE	Native cysteine residues, VC protease-cleavable linker	n/a, 2.5 *mg/kg*, yes	Progenics (74,75)
MLN2704	Phase I & II, for treatment of prostate cancer	hu anti-PSMA (huJ591), n/a, n/a	DM1	Lysine residues, SPP disulfide cleavable linker	n/a, 60 *mg/kg*, yes	Millennium (76)
Targeting CD37 (Tetraspanin-26) antigen, a TP3 present on mature B cells, implicates as a signaling death receptor to regulate B/T-cell interactions/proliferation:						
IMGN529, Naratuximab emtansine	Phase I or II, for treatment of BCL, CLL, NHL	huIgG1anti-CD37 (K7153A), n/a, ADCC and CDC,	DM1	Native lysine residues, SMCC nonreducible thioether linkage	n/a,1.0 *mg/kg*, no	ImmunoGen (78,79)
AGS67E	Phase I, trial for treatment of NHL, DLBCL with high level of CD37 expression	huIgG2κ anti-CD37 (AGS67C or vCD37-9a73), n/a, n/a	MMAE	Native cysteines residues, VC protease-cleavable linker	n/a,1.2 *mg/kg*, yes	Agensys (80–81)
Targeting CD30 (TNFRSF8) antigen, a tumor necrosis factor:						
Adcetris, brentuximab vedotin, SGN-35	Approved in 2011, for treatment of HL and ALL.	Chimeric IgG1anti-CD30 (cAC10 or SGN30), n/a,	MMAE	Native interchain cysteine, MC-VC-PABC linker	~ 4, 1.8 mg/kg, yes	Seattle Genetics (6,7)

Targeting HER3 antigen, a member of EGFR family RTK, frequently overexpressed in solid tumors, including breast, lung, and colorectal tumors of epithelial origin; it has no active kinase domain itself but is activated through heterodimerization with other members of the EGFR family:						
U3-1402	Phase I & II, for treatment of HER3-positive metastatic breast cancer	huIgG1anti-HER3(Patritumab)	DXd	n/a	~8, n/a, n/a	Daiichi Sankyo, Inc. (82)
Targeting DLL3 antigen, scr-like kinase (Fyn3) acts as a notch ligand for cell-cell communication:						
Rovalpituzumab tesirine, Rova-T, SC16LD6.5	Phase I & II, for treatment of SCLC	huIgG1 anti-DLL3 antibody (SC-16), 2.6 *n*M, n/a	PBD	Native inter-chain cysteine, PEG8-va linker, cathepsin-B cleavable dipeptide linker	~ 2, 0.2 *mg/kg*, yes	Stemcentrx (83)
Targeting GPNMB antigen, an EGP is involved in differentiation of osteoblasts, and cellular adhesion:						
Glembatumumab Vedotin (GV), CDX-011, CR011-vcMMAE	Phase II, for treatment of GPNMB-positive breast and melanoma cancer	huIgG2 (CR011), n/a, no	MMAE	Cysteine residues, VC protease-cleavable linker	~ 4.5, 1.9 *mg/kg*, yes	Celldex Therapeutics (84–87)
Targeting CD79b antigen, a TP1 on B cells mediates signal transduction cascade activated by BCR:						
Polatuzumab vedotin, RG7596, DCDS4501A	Phase II, for treatment of NHLs and CLLs	anti-CD79b, n/a, n/a	MMAE	Native cysteine residues, VC protease-cleavable linker	n/a, 2.4 *mg/kg*, yes	Genentech, Inc. (88)
Targeting GCC antigen, a part of calcium negative feedback system and has a role in cGMP synthesizes from GTP:						
Indusatumab vedotin, MLN0264, TAK-264, 5F9-vcMMAE	Phase II, for treatment of GI malignancies	IgG1 anti-GCC (TAK-264), n/a, n/a	MMAE	Native cysteine residues, VC protease-cleavable linker	n/a, ~1.8 *mg/kg*, yes	Millennium (89,90)
Targeting NaPi2b antigen, a sodium phosphate transporter:						
Lifastuzumab vedotin, RG7599, DNIB0600A	Phase II, for treatment of NSCLC and ovarian cancer	huIgG1 anti-NaPi2b, 10.19 *nM*, n/a	MMAE	Native cysteine residues, VC protease-cleavable linker	n/a, 2.4 *mg/kg*, yes	Genentech, Inc. (91,92)
Targeting CA6 antigen, a sialoglycotope of MUC-1 is over-expressed in variety of solid tumors, including breast, ovarian, cervical, lung and pancreatic tumors:						
SAR566658	Phase II, for treatment of OC, breast, cervical, lung cancers	huIgG1 anti-CA6 (huDS6 IgG1), n/a, n/a	DM4	Native lysine residues, SPDB disulfide cleavable linker	6.5 *mg/kg*	Sanofi (93,94)
Targeting CD74 antigen, a TP2 on B cells involved in the formation and transport of MHC class II protein:						
Milatuzumab–doxorubicin, IMMU-110, hLL1-DOX	Phase I & II, for treatment of MM	hu anti-CD74	DOX	Native lysine residues, Acid-labile hydrazone linker	n/a, n/a, yes	Immunomedics (95)
Targeting CD138 antigen, syndecan1, a type I transmembrane heparan sulfate proteoglycan participates in cell proliferation, cell migration and cell-matrix interactions:						
BT-062, Indatuximab ravtansine	Phase I & II, for treatment of MM	Chimeric anti-CD138 (nBT062), n/a, n/a	DM4	Native lysine residues, SPDB disulfide cleavable linker	n/a, 2.7 *mg/kg*, yes	Biotest (96)
Targeting BCMA antigen, a receptor for a proliferation-inducing ligand and B-cell activating factor:						

GSK2857916	Phase I, for treatment of MM	Engineered afucosylated huIgG1 anti-BCMA, 1 *nM*, ADCC	MMAF	Native cysteine residues, MC noncleavable linker	n/a, n/a, no	GlaxoSmithKine (97)
Targeting specific myeloma antigen:						
DFRF4539A, RG7598	Phase I, for treatment of MM	n/a, n/a, n/a	MMAE	n/a	n/a, n/a, n/a	Genentech, Inc. (100)
Targeting SLAMF7 (CS1) antigen:						
ABBV-838	Phase I, for treatment of MM	huIgG1 anti-SLAMF7, n/a, n/a	MMAE	Native cysteine residues, VC protease-cleavable linker	n/a, n/a, n/a	Abbvie (101)
Targeting CD56 antigen, associates with FGFR and stimulates RTKs to induce neurite outgrowth:						
IMGN901, Lorvotuzumab mertansine, huN901-DM1/BB-10901	Phase I & II, for treatment of CD56+ MM	huIgG1 anti-CD56 (Lorvotuzumab or N901), 0.002 nM, ADCC	DM1	Lysine residues, SPP disulfide cleavable linker	3.7, 2.0 *mg/kg*, n/a	ImmunoGen (102)
Targeting ENPP3 (CD203c) antigen, a TP2 belongs to a series of ectoenzymes, possess ATPase and ATP pyrophosphatase activities:						
AGS-16C3F	Phase I & II, for treatment of RRCC	huIgG2k anti-ENPP3 (AGS16-7.8), 0.3–1.1 *nM*, no	MMAF	Native cysteine residues, MC noncleavable linker	~ 4, 1.8 *mg/kg*, no	Astellas Pharma (103,104)
Targeting TF (CD142) antigen, a TP and initiator of the coagulation cascade:						
Humax-TF-ADC, tisotumab vedotin	Phase I & II, for treatment of Multiple solid tumours	IgG1 anti-TF	MMAE	Native cysteine residues, VC protease-cleavable linker	n/a,1.8 *mg/kg*, yes	Genmab (105)
Targeting TIM1 antigen, a member of the T cell transmembrane IgG and mucin family, which plays critical roles in regulating immune cell activity especially regarding the host response to viral infection:						
CDX-014	Phase I & II, for treatment of RCC	huIgG1anti-TIM1	MMAE	Native cysteine residues, VC protease-cleavable linker	n/a, n/a, n/a	Celldex Therapeutics (106)
Targeting FOLR1 antigen, a membrane-bound protein regulates transport of the vitamin B9 into cells:						
IMGN853, mirvetuximab soravtansine	Phase I, for treatment of folate receptor alpha (FRα)-positive cancer, *e.g.*, relapsed EOC	FRa-binding antibody	DM4	Native lysine residues, Sulfo-SPDB disulfide cleavable linker	n/a, 6 *mg/kg*, yes	ImmunoGen (17,107–110)
Targeting MUC16 (CA-125) antigen, a member of the mucin family GP that acts as a lubricating barrier against foreign particles and infectious agents on the apical membrane of epithelial cells:						
RG7458, Sofituzumab Vedotin, DMUC5754A	Phase I, for treatment of ovarian and pancreatic cancer	IgG1anti-MUC16 (OC125), n/a, n/a	MMAE and MMAF	Native cysteine residues, MC-VC-PABC linker	n/a, 2.4 *mg/kg*, yes	Genentech, Inc. (111)
Targeting CanAg antigen, is a novel glycoform of mucin family GP:						
IMGN242, HuC242-DM4, cantuzumab ravtansine	Phase I, for treatment of Non-colorectal and Pancreatic Cancer	hu anti-CanAg (C242 or cantuzumab), n/a, n/a	DM4	Native lysine residues, SPDB disulfide cleavable linker	n/a, n/a, yes	ImmunoGen (112)
Targeting Ckit (CD117 or SCFR) antigen, a TP and RTKs having a key role in the regulation of cell differentiation and proliferation:						

LOP628, Anti c-KIT ADC	Phase I, for treatment of AML and solid tumors	huIgG1anti-(c-Kit), n/a, n/a	DM1	Native lysine residues, SMCC noncleavable thioether linker	n/a, n/a, no	Novartis (113)
Targeting EphA2 antigen, belonging to ephrin receptor subfamily of the RTKs family regulating cell migration, adhesion, proliferation and differentiation:						
MEDI-547, MI-CP177	Phase I, for treatment of relapsed or refractory solid tumors associated with EphA2 expression	huIgG1 anti-EphA2 (1C1), 1nM, n/a	MMAF	Native cysteines residues, MC noncleavable linker	4, 6.0 *mg/kg*, no	Medimmune (114,115)
Targeting Nectin 4 (PVRL4) antigen, a TP1 and member of a family of cellular adhesion molecules, involved in Ca2+-independent cellular adhesion:						
ASG-22ME, AGS-22M6E, anti-nectin-4 ADC, Enfortumab vedotin	Phase I, for treatment of MUC	huIgG1 anti-nectin-4 (AGS-22M6) 0.01 *nM*, n/a	MMAE	Native cysteines residues, VC protease-cleavable linker	n/a,1–3 *mg/kg*, yes	Astellas Pharma (116,117)
Targeting SLTRK6 antigen, belonging to the integral TPs(SLITRK) with LRR:						
AGS15E, anti-SLITRK6 ADC	Phase I, for treatment of MUC	huIgG2γ anti-SLITRK6, n/a, n/a	MMAE	Native cysteines residues, VC protease-cleavable linker	n/a, n/a, yes	Agensys (119)
Targeting HGFR (cMet) antigen, RTKs for hepatocyte growth factor:						
ABBV-399, Telisotuzumab vedotin	Phase I, for treatment of c-Met-expressing NSCLC	Engineered huIgG1 without the agonist activity associated with c-Met (ABT-700), 0.2 to 1.5 *nM*, ADCC and c-Met inhibition & downstream signaling molecules	MMAE	Native cysteines residues, VC protease-cleavable linker	~3.1, 3 *mg/kg*, n/a	Abbvie (120–123)
Targeting FGFR2 antigen, type 2 RTKs with a role in both embryonic development and tissue repair:						
BAY1187982, anti-FGFR2 ADC, Aprutumab ixadotin	Phase I, for treatment of FGFR2-positive human malignancies	huIgG1anti-FGFR2 isoforms FGFR2-IIIb and FGFR2-IIIc (BAY 1179470), 75 *nM*, n/a	MMAE	Lysine side chains and a noncleavable linker	~4, n/a, yes	Bayer (124)
Targeting C4.4a (LYPD3) and uPAR antigen, glycosylphosphatidylinositol (GPI)-anchored proteins:						
BAY1129980, Lupartumab amadotin, anti-C4.4a ADC	Phase I, for treatment of LSCC	huIgG1anti-C4.4A, 60 *nM*, n/a	MMAE	Native cysteine residues, non-cleavable alkyl hydrazide linker	~4, 1.9 *mg/kg*, n/a	Bayer (125)
Targeting p-Cadherin (Cadherin 3) antigen, a cell-surface protein and member of the cadherin family plays a role in cell adhesion, motility, invasion, and proliferation:						
PCA062	Phase I, for treatment of TNBC; head and neck & esophageal cancers	IgG1 anti-P-cadherin, n/a, n/a	DM1	Native lysine residues, SMCC noncleavable thioether linker	n/a, n/a, n/a	Novartis (126)
Targeting 5T4 (TPBG) antigen, a EGP correlated with increased invasiveness:						

PF-06263507, anti-5T4 ADC	Phase I, for treatment of lung and breast cancer with 5T4 expression	huIgG1 anti-5T4	MMAF	Native cysteine residues, MC noncleavable linker	n/a,4.34 *mg/kg*, no	Pfizer (127)
Targeting STEAP1 antigen, cell-surface protein is predominantly expressed in prostate tissue:						
RG7450, DSTP3086S, Vandortuzumab vedotin, STEAP1 ADC	Phase I, for treatment of mCRPC	huIgG1 anti-TEAP1 (MSTP2109A), 2.4 *nM*, n/a	MMAE	Native cysteine residues, MC-vc-PAB linker	1.8–2.0, 2.4 *mg/kg*, yes	Genentech, Inc. (128–131)
Targeting PTK7 antigen, RTKs 7 presents on TICs in the Wnt signaling pathway:						
PF-06647020, h6M24-vc0101, PTK7-targeted ADC	Phase I, for treatment of NSCLC, TNBC and OC	huIgG1anti-PTK7 (h6M24) 0.002 *nM*, n/a	Aur0101	Transglutaminase tag (LLQGA) located at the C-terminus of the antibody heavy chain, cleavable VC-PABC-linker	4, 1.5 *mg/kg*, yes	Pfizer (132,133)
Targeting Ephrin-A4 (EFNA4) antigen, RTKs modulate signaling pathways that impact cell fate decisions during embryogenesis and adult tissue homeostasis:						
PF-06647263	Phase I, for treatment of TNBC and OC	huIgG1anti-Ephrin-A4 (E32), n/a, n/a	Calich.	Native lysine residues, Hydrazone–CM1(Hydrazone acetyl butyrate)	4.6, 0.08 *mg/kg*, yes	Pfizer (113,134)
Targeting LIV1(SLC39A6 or ZIP6) antigen, a member of the zinc transporter family playing a key role in tumor cell progression and metastasis:						
SGN-LIV1A, anti-LIV-1	Phase I, for treatment of metastatic breast,	huIgG1 anti-LIV1(hLIV22), 4.6 *nM*, n/a	MMAE	Native cysteine residues, VC protease-cleavable linker	4, n/a, yes	Seattle Genetics (135)
Targeting TENB2 antigen, a prostate cancer target associated with the progression of poorly differentiated and androgen-independent tumor types:						
Anti-TENB2 ADC	Phase I, for treatment of prostate cancer	ThioMab version of the anti-TENB2 antibody (Pr1), 2.3 *n*M, n/a	MMAE	Native lysine residues, protease-labile VC-PABC-linker	2, n/a, n/a	Seattle Genetics (131,139)
Targeting ETBR antigen, a G-protein coupled receptor that can activate RAF/MEK signaling:						
RG7636, DEDN-6526A	Phase I, for treatment of melanoma	huIgG1 anti-ETBR, n/a, n/a	MMAE	n/a	n/a, 2.4 *mg/kg*, n/a	Genentech, Inc. (140)
Targeting integrin v3 antigen:						
IMGN-388	Phase I, for treatment of NSCLC and prostate cancer	huIgG1anti-Integrin v3	DM4	Native lysine residues, SPDB disulfide cleavable linker	n/a, 3.5 *mg/kg*, n/a	ImmunoGen (141)
Targeting crypto antigen, belonging to the EGF-CFC family of growth factor-like molecules, playing a key role in signaling pathways of certain transforming growth factor-beta super-family members:						
BIIB-015	Phase I, for treatment of breast, ovary, stomach, lung, and pancreas Cripto-expressing tumor cells	huIgG1 anti-Cripto (BIIB015), n/a, n/a	DM4	Native lysine residues, SPDB disulfide cleavable linker	n/a, n/a, n/a	Biogen (142)

Targeting AGS-5 (SLC44A4) antigen, a sodium-dependent transmembrane transport protein:						
ASG-5ME	Phase I, for treatment of pancreatic, prostate and gastric cancers	huIgG2 anti-AGS-5, n/a, n/a	MMAE	Native cysteine residues, VC protease-cleavable linker	n/a, n/a, n/a	Seattle Genetics/ Astellas (143)
Targeting LY6E antigen, an interferon (IFN)-inducible glycosylphosphatidyl inositol (GPI)-linked cell membrane protein:						
RG7841, DLY-E5953A	Phase I, for treatment of HER2– breast cancer and NSCLC	n/a, n/a, n/a	MMAE	Native cysteine residues, VC protease-cleavable linker	n/a, n/a, n/a	Genentech, Inc. (144)
Targeting AXL (UFO) antigen, a member of the TAM (TYRO3, AXL and MER) family of RTK, playing a key role in tumor cell proliferation, survival, invasion and metastasis:						
HuMax-Axl-ADC	Phase I, for treatment of multiple solid tumors	huIgG1anti-AXL, n/a, n/a	MMAE	Native cysteine residues, VC protease-cleavable linker	n/a, n/a, n/a	Genmab (145)
Targeting CD205 antigen, a type I C-type lectin receptor normally expressed on various APC and some leukocyte sub-populations:						
MEN1309/OBT076	Phase I, for treatment of NHL	huIgG1 anti-CD205, n/a, n/a	DM4	Native lysine residues, SPDB disulfide cleavable linker	n/a, n/a, yes	Menarini Ricerche (146)
Targeting CD25 (IL-2R alpha) antigen, a TP and tumor-associated antigen (TAA), expressed on certain cancer cells:						
ADCT-301, anti-CD25-PBD ADC	Phase I, for treatment of AML, ALL, relapsed HL and NHL with CD25-positive	huIgG1against CD25, n/a, n/a	PBD	Cleavable linker	n/a, n/a, n/a	ADC Therapeutics S.A. (147)
Targeting LAMP-1 antigen, playing a key role in cell-cell adhesion and migration:						
SAR428926	Phase I, for treatment of HER2 negative breast expansion in LAMP-1 positive TNBC	huIgG1anti-LAMP1(Ab-1)	DM4	Lysine residues, SPDB	n/a, n/a, n/a	Sanofi (148)
Targeting MN/CA IX antigen, a TGP expressed in some human carcinomas and appears to be involved in cancer cell proliferation and transformation:						
ADC BAY79-4620, MN-IC	n/a	huIgG1 anti-MN/CA IX, n/a, ADCC	MMAE	Native cysteine residues, VC protease-cleavable linker	n/a, n/a, n/a	Bayer (149)

3.6 TARGETING CANCER STEM-LIKE CELLS (CSC)

The plausible involvement of residual tumor cells in tumor recurrence has been unveiled by research and this brings cancer stem cells (CSCs) into the arena. These cells are vitally involved in perpetuating a malignancy and also tumor aggression. The CSC theory of cancer progression elucidates a hierarchy with CSCs at the apex that leads to more number of tumor cells with lower proliferative attributes. CSCs demonstrate the ability to undergo self-renewal and differentiate into different types of cancerous cells to form the predominant portion of a malignancy. These are on the same lines as the attributes of stem cells. Studies have unveiled the expression of stem cell-like markers like CD44, CD133 and CD90 on CSCs with their transplantation in immunodeficient mice resulting in tumor proliferation.

Figure 16 below is illustrative of CSCs in tumorigenesis (Ayob, A.Z., Ramasamy, T.S , 2018: *Open Access: This article is distributed under the terms of the Creative Commons Attribution 4.0 International License (http://creativecommons.org/licenses/by/4.0/), which permits unrestricted use, distribution, and reproduction in any medium, provided you give appropriate credit to the original author(s) and the source, provide a link to the Creative Commons license, and indicate if changes were made. The Creative Commons Public Domain Dedication waiver (http://creativecommons.org/publicdomain/zero/1.0/) applies to the data made available in this article, unless otherwise stated*):

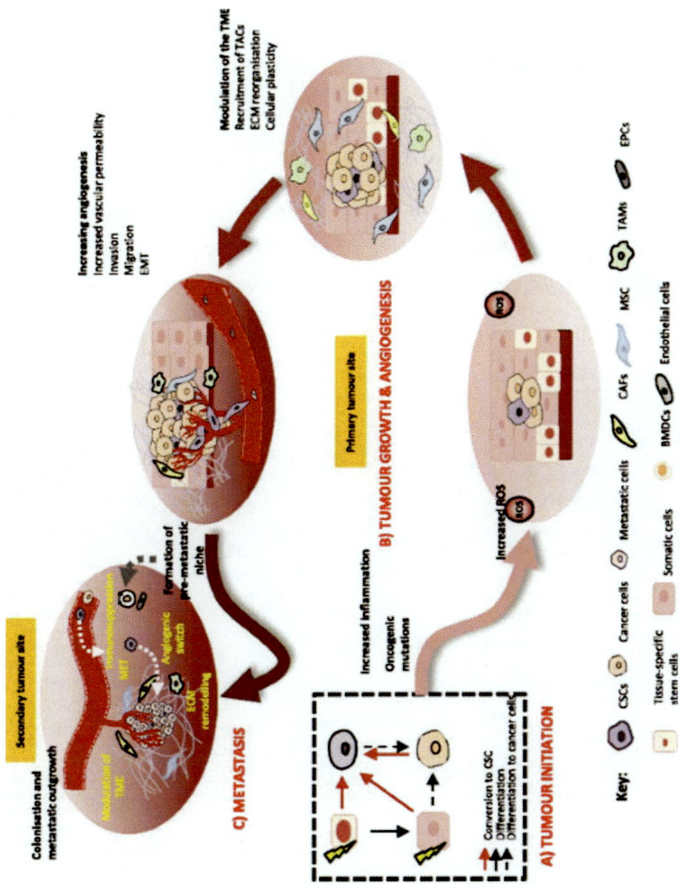

Apart from secreting a slew of signals to "educate" surrounding cells for tumorigenesis, CSCs interact with other vital players like cancer-associated fibroblast (CAFs), endothelial cells, mesenchymal stem cells, macrophages, T-cells and natural killer (NK) cells to facilitate tumor survival.

The roles of CSCs in chemoresistance are also emerging given their quiescent attribute. The other aspect of oncogenesis where CSCs are involved is metastasis due to metastatic CSCs. The entire "cascade" of metastasis entails the interactions of CSCs with other players of the tumor microenvironment for forming new metastatic sites (pre-metastatic niche) for subsequent spread and angiogenesis for secondary tumor formation.

Figure 17 A to C below illustrate the vital involvement of CSCs with crucial stages of oncogenesis:

A) Interactions with the tumor microenvironment for oncogenesis (Ayob, A.Z., Ramasamy, T.S , 2018: *Open Access: This article is distributed under the terms of the Creative Commons Attribution 4.0 International License (http://creativecommons.org/licenses/by/4.0/), which permits unrestricted use, distribution, and reproduction in any medium, provided you give appropriate credit to the original author(s) and the source, provide a link to the Creative Commons license, and indicate if changes were made. The Creative Commons Public Domain Dedication waiver (http:// creativecommons.org/publicdomain/zero/1.0/) applies to the data made available in this article, unless otherwise stated*):

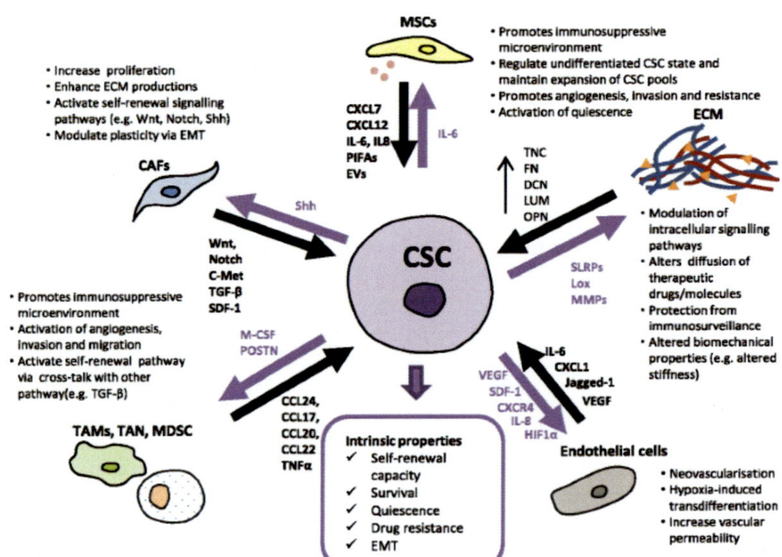

B) ***CSCs and chemoresistance to therapy*** (Ayob, A.Z., Ramasamy, T.S , 2018: *Open Access: This article is distributed under the terms of the Creative Commons Attribution 4.0 International License (http://creativecommons.org/licenses/by/4.0/), which permits unrestricted use, distribution, and reproduction in any medium, provided you give appropriate credit to the original author(s) and the source, provide a link to the Creative Commons license, and indicate if changes were made. The Creative Commons Public Domain Dedication waiver (http://creativecommons.org/publicdomain/zero/1.0/) applies to the data made available in this article, unless otherwise stated*):

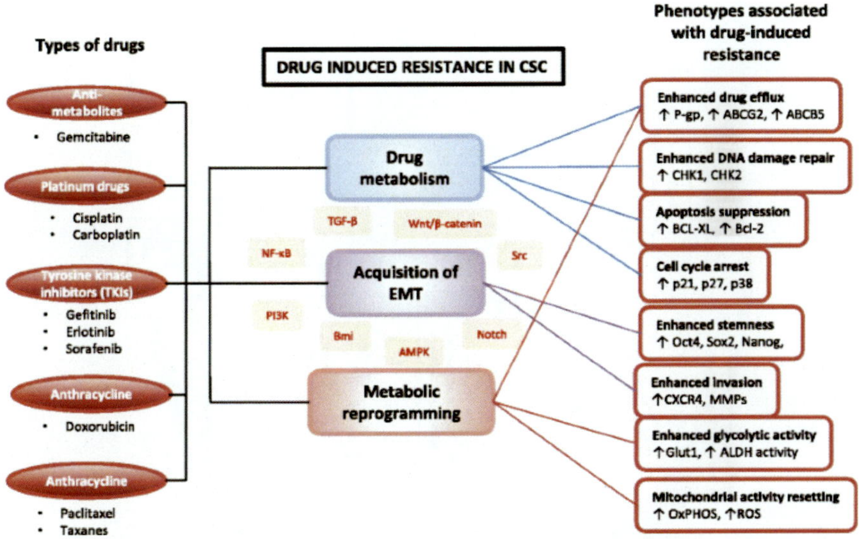

C) ***CSCs and metastasis***: Inducing metastasis by CSCs with Kupffer's cells, endothelial cells, hepatic stellate cells and bone marrow-derived progenitor cells for the pre-metastatic niche establishment. Further growth and angiogenesis are facilitated by CSCs (Ayob, A.Z., Ramasamy, T.S , 2018: *Open Access: This article is distributed under the terms of the Creative Commons Attribution 4.0 International License (http://creativecommons.org/licenses/by/4.0/), which permits unrestricted use, distribution, and reproduction in any medium, provided you give appropriate credit to the original author(s) and the source, provide a link to the Creative Commons license, and indicate if changes were made. The Creative Commons Public Domain Dedication waiver (http://creativecommons.org/publicdomain/zero/1.0/) applies to the data made available in this article, unless otherwise stated*):

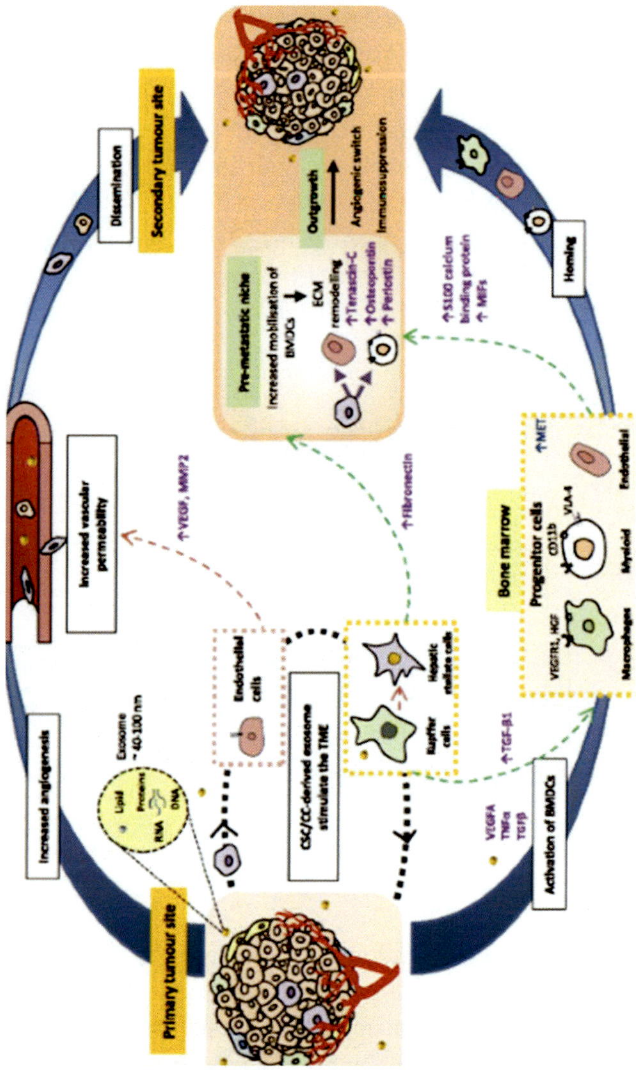

In order to target CSCs, ADCs have entered the stage. An ADC targeting DLL3 expressed more on neuroendocrine carcinoma and small-cell lung cancer (SCLC) called rovalpituzumab tesirine reached phase III in advanced SCLC patients. Studies are in progress for targeting PTK7 on CSCs called PF-06647020: humanized anti-PTK7 mAb, a cleavable dipeptide (valine-citrulline) linker, and Aur0101, an auristatin microtubule inhibitor (Marcucci et al, 2019).

Figure 18 below is illustrative of an ADC targeting CSCs: with a mAb targeting a specific antigen, a linker and a cytotoxic molecule targeting dividing cells (anti-mitotic) or a drug binding DNA to target even quiescent cells: ADC, antibody-drug conjugate; CSC, cancer stem-like cell; mAb, monoclonal antibody (© 2019 Marcucci, Caserta, Romeo and Rumio. *This is an open-access article distributed under the terms of the Creative Commons Attribution License (CC BY). The use, distribution or reproduction in other forums is permitted, provided the original author(s) and the copyright owner(s) are credited and that the original publication in this journal is cited, in accordance with accepted academic practice. No use, distribution or reproduction is permitted which does not comply with these terms*):

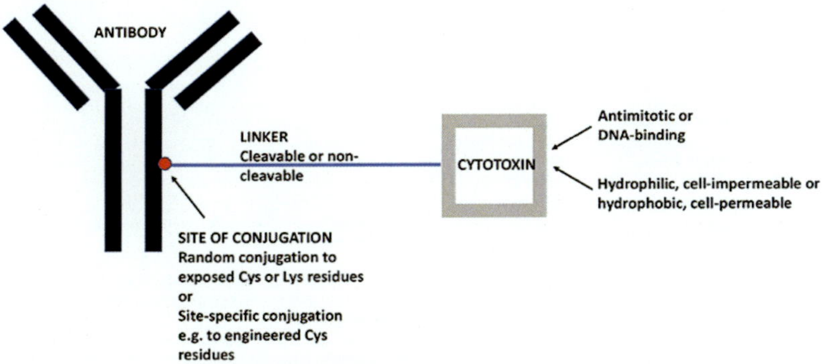

Table 6 below summarizes research on ADCs targeting CSCs (© 2019 Marcucci, Caserta, Romeo and Rumio. *This is an open-access article distributed under the terms of the Creative Commons Attribution License (CC BY). The use, distribution or reproduction in other forums is permitted, provided the original author(s) and the copyright owner(s) are credited and that the original publication in this journal is cited, in accordance with accepted academic practice. No use, distribution or reproduction is permitted which does not comply with these terms*):

Target: ADC name	Linker: Cleavable/uncleavable	Cytotoxin: Antimitotic, DNA binder	Developmental stage Stage, ClinicalTrials.gov identifiers, results if available	References
ADCs AGAINST MARKERS EXPRESSED ON CSC				
DLL3: rovalpituzumab tesirine	Cleavable (dipeptide)	DNA binder (PBD dimer)	Phase III, NCT03061812, vs. topotecan in DLL3^{+} advanced or metastatic SCLC at first disease progression after platinum chemotherapy Phase III, NCT03334487, evaluating the safety for third-line and later treatment of relapsed or refractory SCLC Phase II, NCT02674568, as third-line and later treatment for subjects with relapsed or refractory DLL3^{+} SCLC Other phase I/II studies in patients with DLL3^{+} SCLC (NCT02819999, NCT02874664) or SCLC and other solid tumors (NCT02709889)	(30, 31)
Protein tyrosine kinase 7 (PTK7): PF-06647020	Cleavable (dipeptide)	Antimitotic (Aur0101)	Safety study, NCT03243331: with gedatolisib in TNBC Phase I, NCT02222922: in adult patients with advanced solid tumors	(35)
Ephrin-A4 (EFNA4): PF-06647263	Cleavable (hydrazone)	DNA binder (calicheamicin)	Phase I, NCT02078752: in patients with advanced solid tumors	(36)
IL-3 receptor α chain (CD123): SGN-123	Cleavable (dipeptide)	DNA binder (PBD dimer)	Phase I, NCT02848248, in AML patients. Study terminated, presumably no longer in active development.	(37)
5T4: PF-06263507	Non-cleavable (maleimidocaproyl)	Antimitotic (MMAF)	Phase I, NCT01891669, no objective responses were observed	(38, 39)
5T4: MEDI-0641	Cleavable (dipeptide)	DNA binder (PBD dimer) Antimitotic (tubulysin)	Not reported	(40, 41)
5T4: H6-DM4	Cleavable (SPDB)	Antimitotic (DM4)	Not reported	(42)
LGR5	Cleavable (dipeptide) Non-cleavable (malemidopropionyl)	Antimitotic (MMAE) Antimitotic (MMAE)	Not reported Not reported	(43)
LGR5	Cleavable (dipeptide) Cleavable (acid-sensitive)	Antimitotic (MMAE) DNA binder (PNU159682)	Not reported Not reported	(44)
ANTITUMOR ADCs THAT WERE SHOWN TO HAVE ANTI-CSC ACTIVITY AT A LATER STAGE				
HER2: T-DM1, ado-trastuzumab emtansine	Non-cleavable	Anti-mitotic (DM1)	FDA-approved for the treatment of HER2-positive metastatic breast cancer.	(45)
CD33: gemtuzumab ozogamicin	Cleavable (hydrazone)	DNA binder (calicheamicin)	FDA approval in 2000 for the treatment of AML. Voluntarily withdrawn in 2010 due to safety concerns. Recently reapproved.	(46)
NCAM (CD56): lovortuzumab mertansine	Cleavable	Anti-mitotic (DM1)	It was in development as antitumor agent, not specifically as anti-CSC agent. Development now halted due to disappointing results in lung cancer patients	(47, 48)

ADC, antibody-drug conjugate; AML, acute myeloid leukemia; CSC, cancer stem-like cells; DLL3, Delta-like ligand 3; EFNA4, Ephrin-A4; FDA, Food and Drug administration; HER2, human epidermal growth factor receptor 2; LGR5, leucine-rich repeat-containing, G protein-coupled receptor 5; MMAE, monomethyl auristatin E; NCAM, neural cell-adhesion molecule; PBD, pyrrolobenzodiazepine; PTK7, protein tyrosine kinase 7; SCLC, small-cell lung cancer.

"Cancer may have started the fight, but I will finish it."

— gotCancer.org

3.7 REFERENCES

1. Howlader N, Noone AM, Krapcho M, Miller D, Brest A, Yu M, Ruhl J, Tatalovich Z, Mariotto A, Lewis DR, Chen HS, Feuer EJ, Cronin KA (eds). SEER Cancer Statistics Review, 1975-2018, National Cancer Institute. Bethesda, MD, https://seer.cancer.gov/csr/1975_2018/, based on November 2020 SEER data submission, posted to the SEER web site, April 2021.
2. Baah S, Laws M, Rahman KM. Antibody–Drug Conjugates—A Tutorial Review. *Molecules*. 2021; 26(10):2943. https://doi.org/10.3390/molecules26102943
3. Hafeez, U., Parakh, S., Gan, H. K., & Scott, A. M. (2020). Antibody-Drug Conjugates for Cancer Therapy. *Molecules (Basel, Switzerland)*, *25*(20), 4764. https://doi.org/10.3390/molecules25204764
4. Ponziani, S., Di Vittorio, G., Pitari, G., Cimini, A. M., Ardini, M., Gentile, R., Iacobelli, S., Sala, G., Capone, E., Flavell, D. J., Ippoliti, R., & Giansanti, F. (2020). Antibody-Drug Conjugates: The New Frontier of Chemotherapy. *International journal of molecular sciences*, *21*(15), 5510. https://doi.org/10.3390/ijms21155510
5. Nejadmoghaddam, M. R., Minai-Tehrani, A., Ghahremanzadeh, R., Mahmoudi, M., Dinarvand, R., & Zarnani, A. H. (2019). Antibody-Drug Conjugates: Possibilities and Challenges. *Avicenna journal of medical biotechnology*, *11*(1), 3–23.
6. Chalouni, C., Doll, S. Fate of Antibody-Drug Conjugates in Cancer Cells. *J Exp Clin Cancer Res* 37, 20 (2018). https://doi.org/10.1186/s13046-017-0667-1
7. Ayob, A.Z., Ramasamy, T.S. Cancer stem cells as key drivers of tumour progression. *J Biomed Sci* 25, 20 (2018). https://doi.org/10.1186/s12929-018-0426-4
8. Marcucci F, Caserta CA, Romeo E and Rumio C (2019) Antibody-Drug Conjugates (ADC) Against Cancer Stem-Like Cells (CSC)—Is There Still Room for Optimism? *Front. Oncol.* 9:167. doi: 10.3389/fonc.2019.00167

CHAPTER 4

ADCs Continued-what's New

"Research is creating new knowledge."

- Neil Armstrong

Contents

4.1 BACKGROUND

As of now, nine ADCs have received FDA approval for two solid tumors and six hematological malignancies. They are summarized in table 1 below (Hammood M, Craig AW, Leyton JV (2021): © 2021 by the authors. Licensee MDPI, Basel, Switzerland. *This article is an open access article distributed under the terms and conditions of the Creative Commons Attribution (CC BY) license: https://creativecommons.org/licenses/by/4.0/*):

Key: [1] *For adults.* [2] In addition, for relapsed or refractory AML in adults and in pediatric patients 2-years and older; [3] Programmed death receptor-1 and programmed death-ligand 1

ADC	**Target**	**Indications and Usage** [1]
Gemtuzumab ozogamicin (GO)	CD33	• Newly diagnosed and relapsed and refractory acute myeloid leukemia (AML) [2]
Brentuximab vedotin (BV)	CD30	• Hodgkin lymphoma after failure of autologous stem cell transplant (ASCT) or after failure of at least two prior multi-agent chemotherapy regimens in patients who are not eligible for ASCT • Systemic anaplastic large cell lymphoma (ALCL) after failure of at least one prior multi-agent chemotherapy regimen
Inotuzumab ozogamicin (InO)	CD22	• Relapsed or refractory B cell precursor acute lymphoblastic lymphoma (ALL)
Polatuzumab vedotin (PV)	CD79b	• In combination with bendamustine and rituximab for relapsed or refractory diffuse large B cell lymphoma after at least two prior treatments
Sacituzumab govitecan (SG)	Trop2	• Triple-negative breast cancer after at least two prior therapies for metastatic disease
Trastuzumab emtansine (T-DM1)	HER2	• Metastatic breast cancer patients who previously received trastuzumab and a taxane • Adjuvant treatment in early breast cancer with residual invasive disease after neoadjuvant taxane- and trastuzumab-based treatment
Trastuzumab deruxtecan (T-DXd)	HER2	• Unresectable or metastatic breast cancer after two or more prior anti-HER2-based regimens in the metastatic setting
Enfortumab vedotin (EV)	Nectin-4	• Locally advanced or metastatic urothelial cancer after a PD-1 [3] or PD-L1 inhibitor, and platinum-containing chemotherapy in the neoadjuvant/adjuvant, locally, advanced or metastatic setting
Belantamab mafodotin (BM)	BCMA	• Relapsed or refractory multiple myeloma after at least four prior therapies including an anti-CD38 mAb, a proteasome inhibitor, and an immunomodulatory agent

A combination of a linker and a potent cytotoxic payload led to ADC development revolving around these two aspects. First and second-generation ADCs are depicted in figure 1 below (Joubert N, Beck A, Dumontet C, Denevault-Sabourin C, 2020: © 2020 by the authors. Licensee MDPI, Basel, Switzerland. *This article is an open access article distributed under the terms and conditions of the Creative Commons Attribution (CC BY) license: http://creativecommons.org/licenses/by/4.0/*):

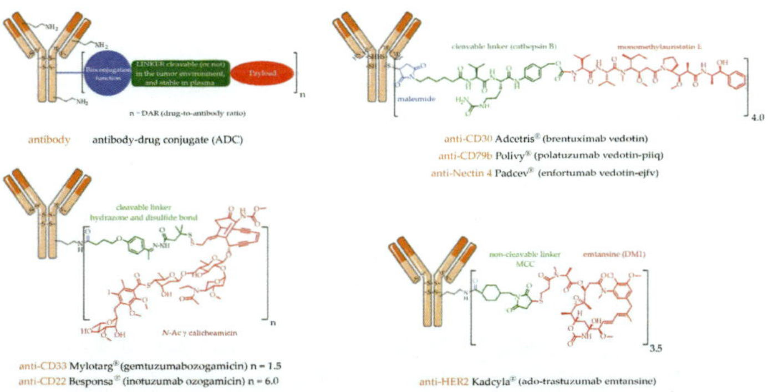

A slew of factors enter the scene for ADCs that need to be kept in mind for effective translation into the clinic as illustrated exhaustively in figure 2 below (© 2019 Tang, Liu, Yu, Sun, Lin, Liu, Han, Wei and Jin. *This is an open-access article distributed under the terms of the Creative Commons Attribution License (CC BY). The use, distribution or reproduction in other forums is permitted, provided the original author(s) and the copyright owner(s) are credited and that the original publication in this journal is cited, in accordance with accepted academic practice. No use, distribution or reproduction is permitted which does not comply with these terms*):

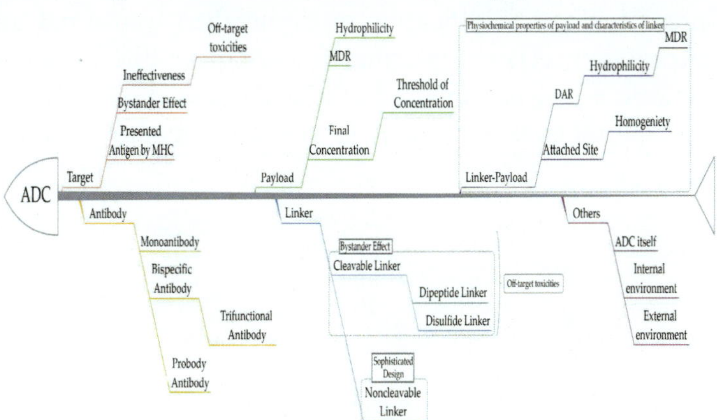

4.2 OVERCOMING ADC RESISTANCE

ADCs too document resistance as an issue on the same lines as other approaches in cancer therapy. This can entail antigen mutations or antigen downregulation to diminish its recognition by the mAb. An example is a patient with anaplastic large-cell lymphoma documenting resistance to brentuximab vedotin due to CD 30 antigen expression loss. Another point is the occurrence of ligands: cell lines have documented lowered T-DM1 (HER2-targeted antibody–drug conjugate trastuzumab emtansine) activity due to the presence of the ErbB3 ligand neuregulin. As elucidated in the previous chapter, the process of clathrin- or caveolae-mediated endocytosis is vital for ADC internalization. Alterations in the trafficking pathway or defects in the proteolytic enzymes involved in processing can create a hurdle here. Another cause of resistance is inhibition of the payload from being released on account of aberrant lysosomal transporter expression.

Altered cell cycle functioning and drug efflux pumps also function in resistance. Aberrant apoptotic pathways also are involved with gemtuzumab ozogamicin resistance arising due to elevated BCl-2 and Bcl-x (anti-apoptotic) and lowered Bak and Bax (proapoptotic) activities (Hafeez, U., Parakh, S., Gan, H. K., & Scott, A. M, 2020).

Figure 3 below summarizes the aspects of resistance to ADCs (Hafeez, U., Parakh, S., Gan, H. K., & Scott, A. M, 2020: © 2020 by the authors: Licensee MDPI, Basel, Switzerland. *This article is an open access article distributed under the terms and conditions of the Creative Commons Attribution (CC BY) license: http://creativecommons.org/licenses/by/4.0/*):

- Antigen loss or mutation; (B) altered internalization pathway and reduced cell surface trafficking; (C) impaired lysosomal proteolysis or acidification or diminished lysosomal transporter expression; (D) cytoplasmic factors, polo-like kinase 1 halting mitotic arrest or impaired cyclin B1 induction; (E) drug efflux transporter overexpression; (F) aberrant apoptotic pathway.

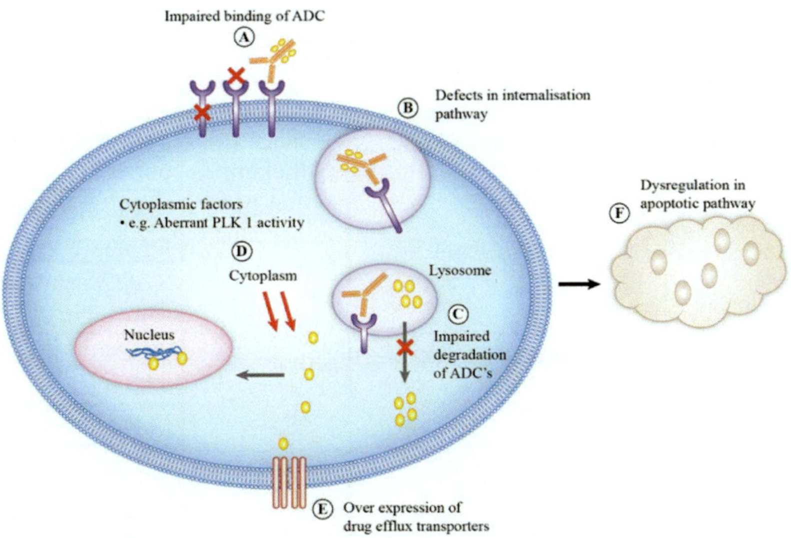

Table 2 below documents the mechanisms of resistance against ADCs that have received FDA approval as unearthed by research (Collins, D. M., Bossenmaier, B., Kollmorgen, G., & Niederfellner, G. (2019): © 2019 by the authors: Licensee MDPI, Basel, Switzerland. *This article is an open access article distributed under the terms and conditions of the Creative Commons Attribution (CC BY) license: http://creativecommons.org/licenses/by/4.0/*):

Key: Complex; Bak = Bcl-2 antagonist/killer; *Bax* = Bcl-2-associated x protein; MMAE = *MonoMethyl Auristatin E; HER2 = Human Epidermal growth factor Receptor 2; SLC46A3 = SoLute Carrier 46A3; STAT3 = Signal Transducer and Activator of Transcription 3; PTEN = Phosphatase and TENsin homolog; V-ATPase = Vacuolar-type proton pumping Adenosine Tri-Phosphate hydrolyzing enzyme; PLK1 = Polo-Like Kinase 1*

ADC	**Resistance Mechanisms Directed Against**		
	Targeting Moiety	**Linker**	**Payload**
Gemtuzumab ozogamicin	CD33 splice variant lacking antibody epitope		drug efflux transporters,
			PI3K pathway activation,
			mTORC1/2 activation,
			deficient Bak/Bax activation

Brentuximab vedotin	CD30 down-regulation		drug efflux transporter,
			MMAE resistance
Trastuzumab emtansine	HER2 down-regulation,	enhanced trafficking to non-lysosomal compartments, reduced V-ATPase activity *	drug efflux transporter,
			SLC46A3 down-regulation,
			STAT3 pathway activation,
	altered internalization		PTEN/PI3K activation,
			PLK1 activation,
			failure to induce Cyclin B1
Inotuzumab ozogamicin			drug efflux transporters

Given the complexity associated with ADCs, a slew of escape routes are available to tumors. Replacing the linker-payload of HER2-targeted antibody–drug conjugate trastuzumab emtansine (T-DM1) with a membrane-permeable payload documented success in preclinical studies. The effect of this bystander killing in patients remains to be gauged. Another interesting approach is boosting the delivery of the payload to the cytosol by a patent by Ward, Ober, Kang, Sun W (2018). This entails an early dissociation of the ADC from its target by engineering the ADC to have a lower affinity in the endolysosomal surrounding over the extracellular space (pH < 6.5; $[Ca^{2+}] \sim 2$ μM) vs. pH > 6.8; $[Ca^{2+}] \sim 2$ mM).

Figure 4 below depicts ADCs and the "escape" of tumors (Collins, D. M., Bossenmaier, B., Kollmorgen, G., & Niederfellner, G. (2019): © 2019 by the authors: Licensee MDPI, Basel, Switzerland. *This article is an open access article distributed under the terms and conditions of the Creative Commons Attribution (CC BY) license: http://creativecommons.org/licenses/by/4.0/*):

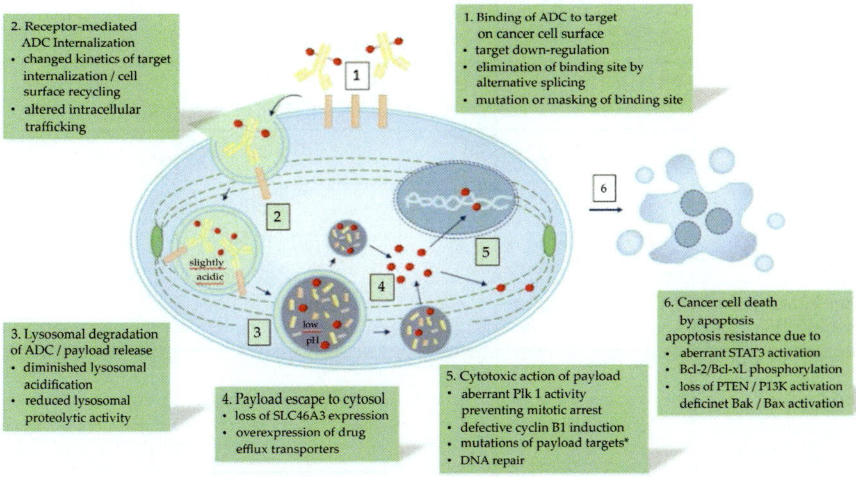

Selected combination therapies are also demonstrating promise in terms of quelling resistance to ADCs. This can entail the use of ADCs with cytotoxic agents like immunotoxins and antibody targeted amanitin conjugates (ATACs) or with immune checkpoint inhibiting (ICI) antibodies. A list of studies is presented below:

Table 3: ADCs with other chemotherapeutic molecules (Collins, D. M., Bossenmaier, B., Kollmorgen, G., & Niederfellner, G. (2019): © 2019 by the authors: Licensee MDPI, Basel, Switzerland. *This article is an open access article distributed under the terms and conditions of the Creative Commons Attribution (CC BY) license: http://creativecommons.org/licenses/by/4.0/*):

Key: GO = Gemtuzumab ozogamicin (Mylotarg ™), T-DM1 = trastuzumab emtansine (Kadcyla ™), INO = Inotuzumab ozogamicin (Besponsa ™), MoA = Mechanism of Action, G-CSF/G-CSFR = Granulocyte–Colony-Stimulating Factor and its Receptor, AML = Acute Myeloid Leukemia, ASCT = Autologous Stem Cell Transplant, ALL = Acute Lymphoblastic Leukemia, CML = Chronic Myeloid Leukemia, DLBCL = Diffuse Large B-Cell Lymphoma, [1] all-trans retinoic acid, [2] nucleophosmin-1, [3] relapsed/refractory

Combination	Trial Number	Phase	MoA of Combined Agent(s)	Indication
GO and 5-azacitidine	NCT00766116	I/II	DNA methyl-transferase inhibitor	Relapsed AML
GO and the combo of idarubicin, etoposide, cytarabine, pegfilgrastim with ATRA [1]	NCT00893399	III	Three different DNA damaging agents plus agonists for G-CSF and retinoic acid receptor	AML with NPM1 mutation [2]
GO and glasdegib	NCT03390296	Ib/II	Smoothened inhibitor	RR [3] AML
Conditioning therapy with GO plus cyclophosphamide and busulfan chemotherapy followed by ASCT	NCT02221310	II	Two immune suppressive alkylating agents	High-risk AML or myelodysplastic syndrome (MDS)
GO and G-CSF, cladribine, cytarabine, and mitoxantrone	NCT03531918	I/II	Three different DNA damaging agents plus a G-CSFR agonist	1st line AML
GO and daunorubicin/cytarabine filled liposomes	NCT03672539	I	Two different DNA damaging agents	RR [3] AML, high-risk MDS
Neoadjuvant T-DM1 and lapatinib followed by Abraxane	NCT02073487	II	A HER1/2 kinase and a microtubule inhibitor	$HER2^+$ breast cancer
T-DM1 and poziotinib	NCT03429101	I	Covalent HER1/2/4 kinase inhibitor	Metastatic $HER2^+$ breast cancer
T-DM1 and osimertinib	NCT03784599	II	HER1 T790M kinase inhibitor	Mutant $HER1^+$, $HER2^+$, stage IV lung cancer
T-DM1 and neratinib	NCT02236000	Ib/II	Irreversible pan-HER inhibitor	Metastatic $HER2^+$ breast cancer
T-DM1 and palbociclib	NCT01976169	Ib	CDK4/6 inhibitor	Recurrent or metastatic HER2+ breast cancer
T-DM1 and ribociclib	NCT02657343	Ib/II	CDK4/6 inhibitor	Metastatic $HER2^+$ breast cancer
T-DM1 and palbociclib	NCT03530696	II	CDK4/6 inhibitor	Metastatic $HER2^+$ breast cancer
T-DM1 and taselisib	NCT02390427	Ib	Phosphoinositide 3-kinase α inhibitor	Advanced $HER2^+$ breast cancer

INO and cyclophosphamide, vincristine, prednisone	NCT01925131	I	An alkylating agent, a microtubule inhibitor, and a glucocorticoid	RR [3] CD22⁺ acute leukemia
INO and low dose chemotherapy (cyclophosphamide/vincristine or methotrexate/cytarabine)	NCT01371630	I/II	Well tolerated cytostatic agents	Older patients with previously untreated ALL
INO and bosutinib	NCT02311998	I/II	Bcr-Abl kinase inhibitor	CD22 and Philadelphia-chromosome positive ALL and CML
INO and intensive chemo-therapy (Hyper-CVAD regimen)	NCT03488225	II	11 induction therapy plus two maintenance therapy agents	1st line B-cell lineage ALL
INO and rituximab, cyclophosphamide, vincristine, prednisolone	NCT01679119	II	Anti-CD20 antibody, well tolerated cytostatic agents	DLBCL patients unfit for anthracycline

Table 4: ADCs with immune checkpoint inhibitors (Collins, D. M., Bossenmaier, B., Kollmorgen, G., & Niederfellner, G. (2019): © 2019 by the authors: Licensee MDPI, Basel, Switzerland. *This article is an open access article distributed under the terms and conditions of the Creative Commons Attribution (CC BY) license: http://creativecommons.org/licenses/by/4.0/*):

Key: T-DM1 = trastuzumab emtansine (Kadcyla ™), BV = Brentuximab vedotin (Adcetris ™), MoA = Mechanism of Action, PD-L1 = Programmed cell Death-Ligand 1, CTLA-4 = Cytotoxic T-Lymphocyte-Associated Protein 4 [1] relapsed/refractory

Combination	**Trial Number**	**Phase**	**MoA of Combo Agent(s)**	**Indication**
T-DM1 and pembrolizumab	NCT03032107	Ib	PD-1 blocking antibody	Metastatic HER2⁺ breast cancer
T-DM1 and atezolizumab	NCT02924883	Ib	PD-L1 blocking antibody	Locally advanced or metastatic HER2⁺ breast cancer
Different doses of T-DM1 and atezolizumab	NCT02605915	Ib	PD-L1 blocking antibody	Locally advanced or metastatic HER2⁺ breast cancer

T-DM1 and utomilumab	NCT03364348	Ib	Agonistic 4-1BB antibody	HER2$^+$ advanced breast cancer
BV and nivolumab	NCT02581631	I/II	PD-1 blocking antibody	RR [1] CD30$^+$ Hodgkin Lymphoma
BV and pembrolizumab	NCT02684292	III	PD-1 blocking antibody	RR [1] CD30$^+$ Hodgkin Lymphoma
BV and nivolumab +/− ipilimumab	NCT01896999	I/II	PD-1 and CTLA-4 blocking antibodies	RR [1] CD30$^+$ Hodgkin Lymphoma

4.3 CASE STUDY: ADC PLUS IMMUNE CHECKPOINT INHIBITORS (ICIS):

Researchers Iwata TN, Sugihara K, Wada T, Agatsuma T reported the benefits of [fam-] trastuzumab deruxtecan with immune checkpoint inhibitors (ICIs) in a mouse model. The antitumor functioning of the ADC [Fam-] trastuzumab deruxtecan (DS-8201a) targeting HER2 made of a humanized anti-human HER2 (anti-hHER2) antibody, an enzymatically-cleavable peptide-based linker, and an exatecan derivative topoisomerase I inhibitor has been documented in a slew of malignancies. Administration of this ADC with anti–CTLA-4 antibody (an ICI) documented a "superior" antitumor profile (a complete response or CR=85%) over [fam-] trastuzumab deruxtecan (10 mg/kg, once a week, twice) or anti–CTLA-4 antibody (5 mg/kg, twice a week, 2 cycles) alone.

Figure 5 below depicts the outcome of DS-8201a and anti–CTLA-4 antibody combinatorial therapy in a EMT6-hHER2 syngeneic mouse model. **A** Mean tumor volumes and standard errors (n = 10 in the vehicle and monotherapy groups; n = 20 in the combination group: red; DS-8201a, blue; anti–CTLA-4 antibody). **B** Survival curve of the same experiment (© 2019 Iwata et al. *This is an open access article distributed under the terms of the Creative Commons Attribution License, which permits unrestricted use, distribution, and reproduction in any medium, provided the original author and source are credited*):

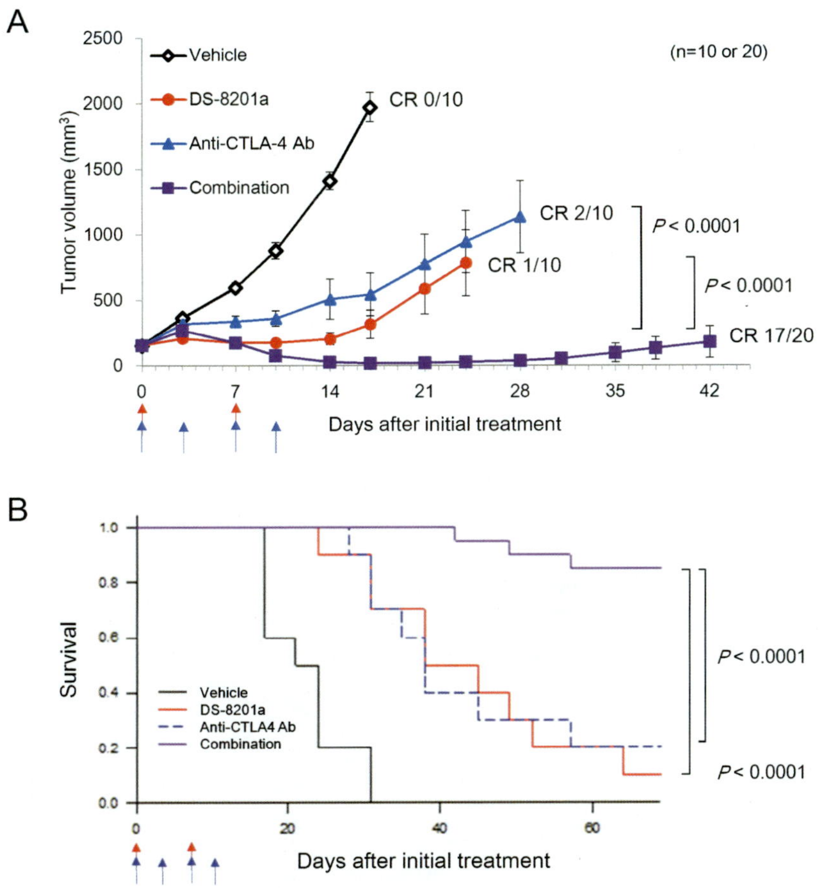

Further analyses unveiled elevated CD45⁺ cells, CD4 and CD8⁺ T cell profiles suggestive of higher intra-tumor T cells. The development of immune memory was also documented by challenging the CR mice and naïve mice with EMT6-hHER2 or EMT6-mock cells; the splenocytes of the former could elicit IFN-γ (immune memory) vs. the latter. Detailed elucidation can further open the avenue of employing such combinatorial approaches of [fam-] trastuzumab deruxtecan with ICIs for more anti-tumor activity and long-term effects.

Figure 6 below is illustrative of these observations. ***TOP***: A) Higher CD45+ cells among live cells; B) Percentages of CD8+ T cells (CD45+ TCRβ+ CD8+ cells) among live cells; C) Percentages of CD4+ T cells (CD45+ TCRβ+ CD4+ cells) among live cells. The circles show individual data and the bars show means with standard errors (n = 6). Dunnett's multiple comparison test: **P < 0.01 and ***P < 0.001. Jonckheere's trend tests: ##P < 0.01 and ###P < 0.001 among the vehicle, DS-8201a, and combination groups, $$$P < 0.01$ and $$$$P < 0.001$ among the vehicle, anti-CTLA-4 antibody, and combination groups; ***BOTTOM***: Anti-tumor activity of ADC + ICI therapy in mice; A, Mean tumor volumes and standard errors (n = 8); B, Spider plots of individual data (© 2019 Iwata et al. *This is an open access article distributed under the terms of the Creative Commons Attribution License, which permits unrestricted use, distribution, and reproduction in any medium, provided the original author and source are credited*):

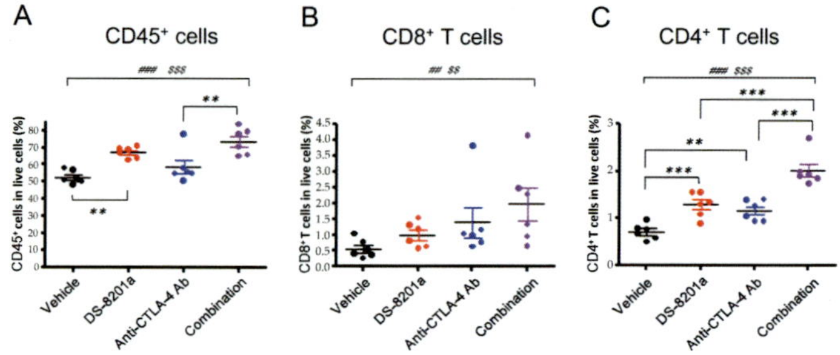

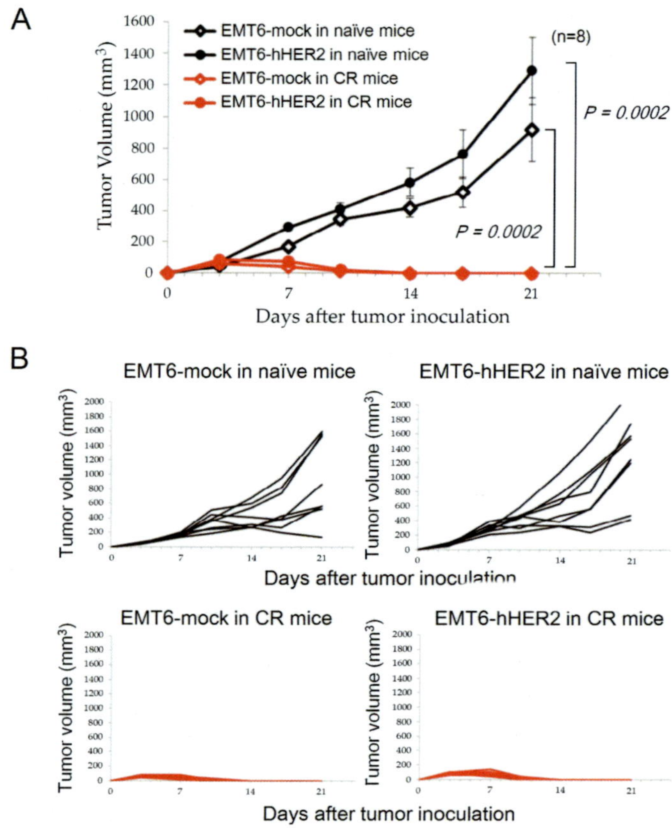

4.4 CASE STUDY: DUAL DRUG ADCS

2021-published research in *Nature Communications* by Yamazaki and team demonstrated the promise of homogeneous ADCs with two distinct payloads to address the intratumor heterogeneity in HER2 heterogeneous tumors challenging the functioning of ADCs. The drugs employed were monomethyl auristatin E (MMAE: cell membrane-permeable and is known for the bystander effect) and monomethyl auristatin F (MMAF: cell membrane-impermeable and cannot be eliminated by the cell once internalized).

Synthesis of the ADRs entailed chemoenzymatic conjugation at specified drug-to-antibody ratios (DARs) followed by characterization. Cytotoxicity probing in HER2-positive (KPL-4, JIMT-1, and SKBR-3) and HER2-negative (MDA-MB-231) breast cancer cell lines, human embryonic kidney 293 (HEK293) cells, and human hepatocyte carcinoma (HepG2) cells revealed toxicity only in the HER2 positive lines. Analyses in healthy BALB/c mice revealed no significant loss of payload with the least toxicity.

Scrutiny in HER2-positive human breast xenografts in mice revealed suppression of tumor growth by the MMAE/F 4+2 dual-drug ADC with no acute toxic effects on the animals. The researchers stated that "*In particular, our data demonstrate the advantages of dual-drug ADCs for treating low-HER2 breast tumors over co-administration of two single-drug ADCs carrying the same payloads, which one may think would be as effective as the former approach*". Further detailed scrutiny can pave the way for such a system to address ADC resistance.

Figure 7 below documents the promise of the initial study of a dual-drug ADC for HER2+ breast cancer xenografts: ***TOP:*** Generating the dual-drug ADCs (Yamazaki et al, 2021: *Open Access: This article is licensed under a Creative Commons Attribution 4.0 International License, which permits use, sharing, adaptation, distribution and reproduction in any medium or format, as long as you give appropriate credit to the original author(s) and the source, provide a link to the Creative Commons license, and indicate if changes were made. The images or other third party material in this article are included in the article's Creative Commons license, unless indicated otherwise in a credit line to the material. If material is not included in the article's Creative Commons license and your intended use is not permitted by statutory regulation or exceeds the permitted use, you will need to obtain permission directly from the copyright holder. To view a copy of this license, visit http://creativecommons.org/licenses/by/4.0/*):

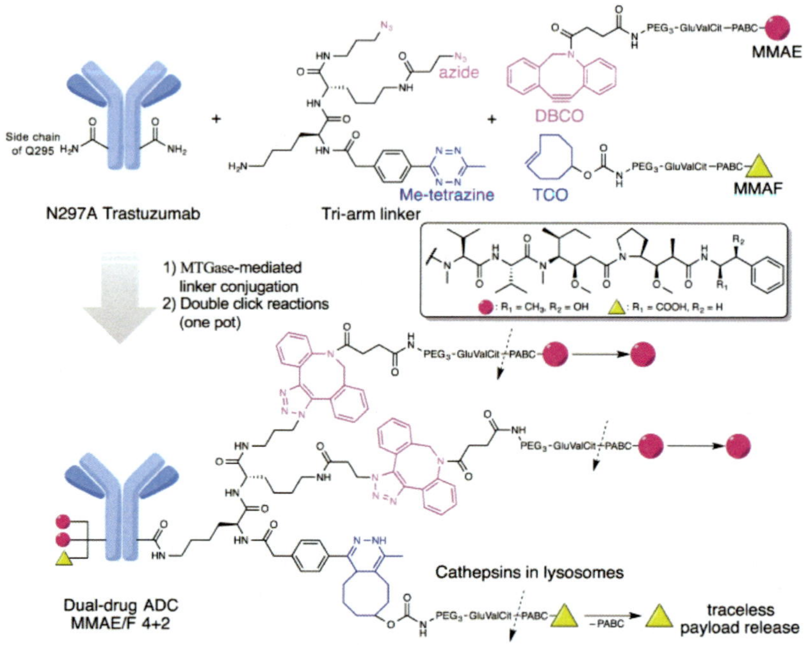

BOTTOM: ADCs in xenograft mice administered a single dose of a 1 : 1 mixture of MMAF DAR 4 and MMAE DAR 4 ADCs (magenta open triangle, $1\,mg\,kg^{-1}$ each), MMAE/F 4+2 ADC (cyan square, $1\,mg\,kg^{-1}$), or vehicle control (black circle). **b** Tumors collected after 24 days of injection. Scale bar: 1 cm. **c** Tumor weights at Day 24. Lines depict mean values (Yamazaki et al, 2021: *Open Access: This article is licensed under a Creative Commons Attribution 4.0 International License, which permits use, sharing, adaptation, distribution and reproduction in any medium or format, as long as you give appropriate credit to the original author(s) and the source, provide a link to the Creative Commons license, and indicate if changes were made. The images or other third party material in this article are included in the article's Creative Commons license, unless indicated otherwise in a credit line to the material. If material is not included in the article's Creative Commons license and your intended use is not permitted by statutory regulation or exceeds the permitted use, you will need to obtain permission directly from the copyright holder. To view a copy of this license, visit http://creativecommons.org/licenses/by/4.0/*):

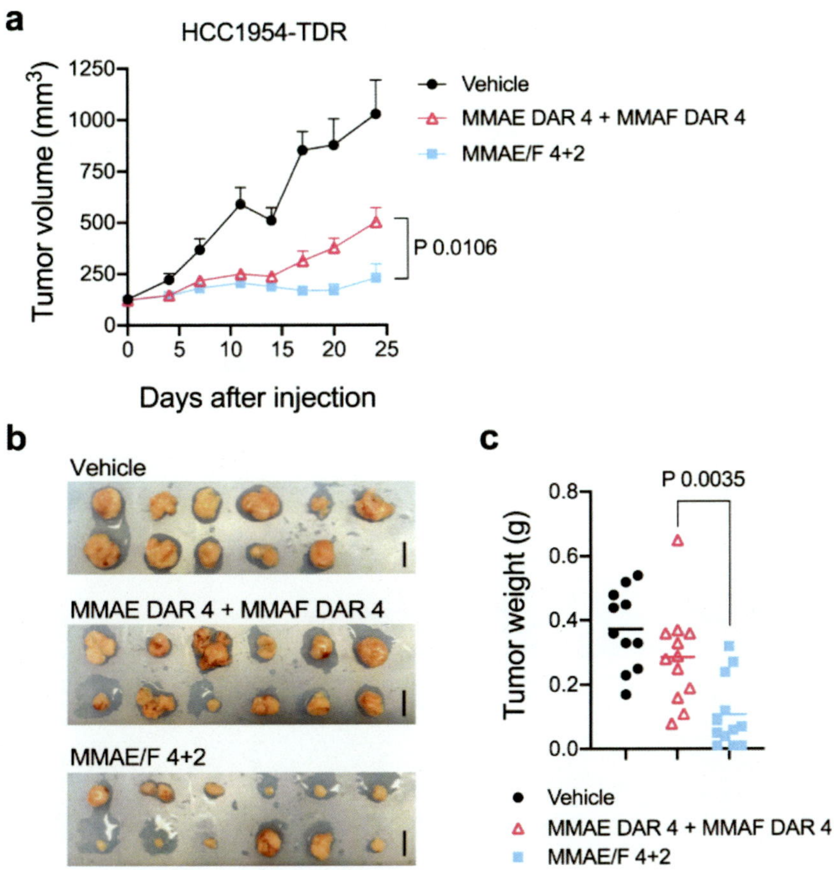

4.5 LINKER CONJUGATION

The stability of the linker in an ADC emerges as essential as the maximum ADC level needs to be maintained while diminishing any payload release in the blood. An acid-labile hydrazone linker is present in gemtuzumab ozogamicin (Mylotarg™) and inotuzumab ozogamicin (Besponsa™) that is stable at the blood pH (7.4) and is hydrolyzed in the lysosome. A challenge here is the lower stability over other linkers. The high reduced glutathione in the cell reduces the non-enzymatic cleavable disulfide linker: this is present in gemtuzumab ozogamicin (Mylotarg™), inotuzumab ozogamicin (Besponsa™) [these two have both hydrazone and disulfide linkers], mirvetuximab soravtansine and anetumab ravtansine.

In addition to the enzyme-cleavable valine-citrulline (Val-Cit) dipeptide linkages in a slew of ADCs like rentuximab vedotin, polatuzumab vedotin, enfortumab vedotin, and glembatumumab vedotin, the use of a tetrapeptide linker [Gly-Gly-Phe-Gly (GGFG)] has been documented in trastuzumab deruxtecan (Shim H, 2020).

Figure 8 below depicts ACS linkers (a) Non-cleavable thioether linker in trastuzumab emtansine (b) Acid-labile hydrazone and disulfide linkage linkers in gemtuzumab ozogamicin and inotuzumab ozogamicin (c) Enzyme-cleavable Val-Cit linker in brentuximab vedotin : Shim H (2020) (© 2020 by the author.Licensee MDPI, Basel, Switzerland. *This article is an open access article distributed under the terms and conditions of the Creative Commons Attribution (CC BY) license: http://creativecommons.org/licenses/by/4.0/*):

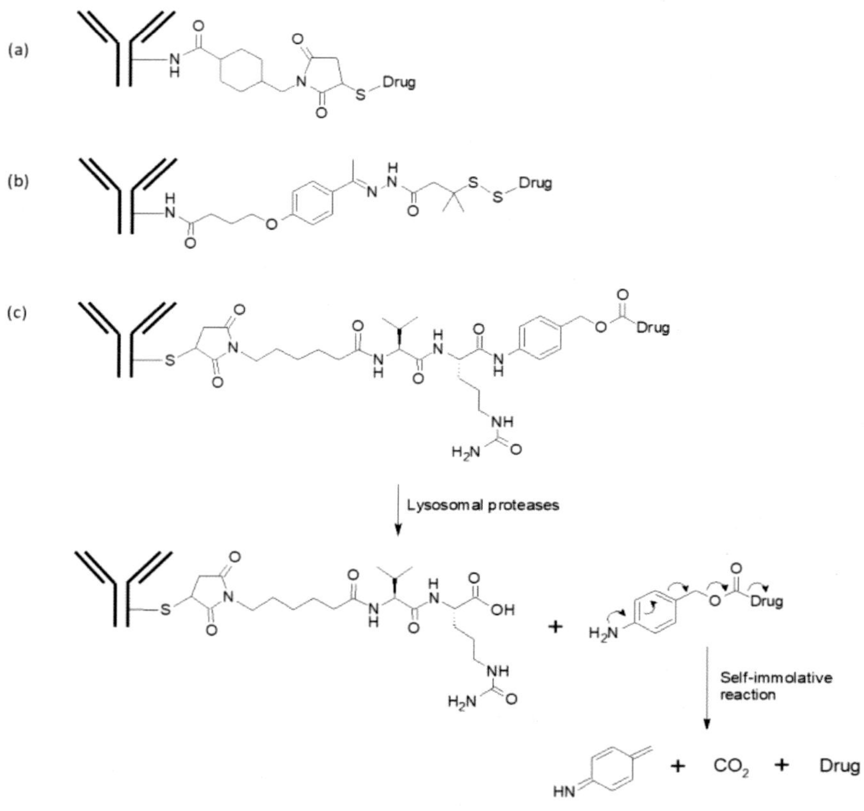

The cysteine and lysine residues have been used to conjugate the linkers to antibodies. Stable conjugates are the outcome of maleimide and alkyl-halide with thiol groups. In order to augment the stability of such an approach, earlier payload loss due to the retro-Michael reaction of thiomaleimide is quelled by semi-hydrolysis of maleimide. Other systems include the engineering of amino acid side chains to form hydrazones or imines or amide formation from lysines.

Figure 9 below is representative of the approaches for linker chemistry (Leung D, Wurst JM, Liu T, Martinez RM, Datta-Mannan A, Feng Y, 2020: © 2020 by the authors. Licensee MDPI, Basel, Switzerland. *This article is an open access article distributed under the terms and conditions of the Creative Commons Attribution (CC BY) license: http://creativecommons.org/licenses/by/4.0/*):

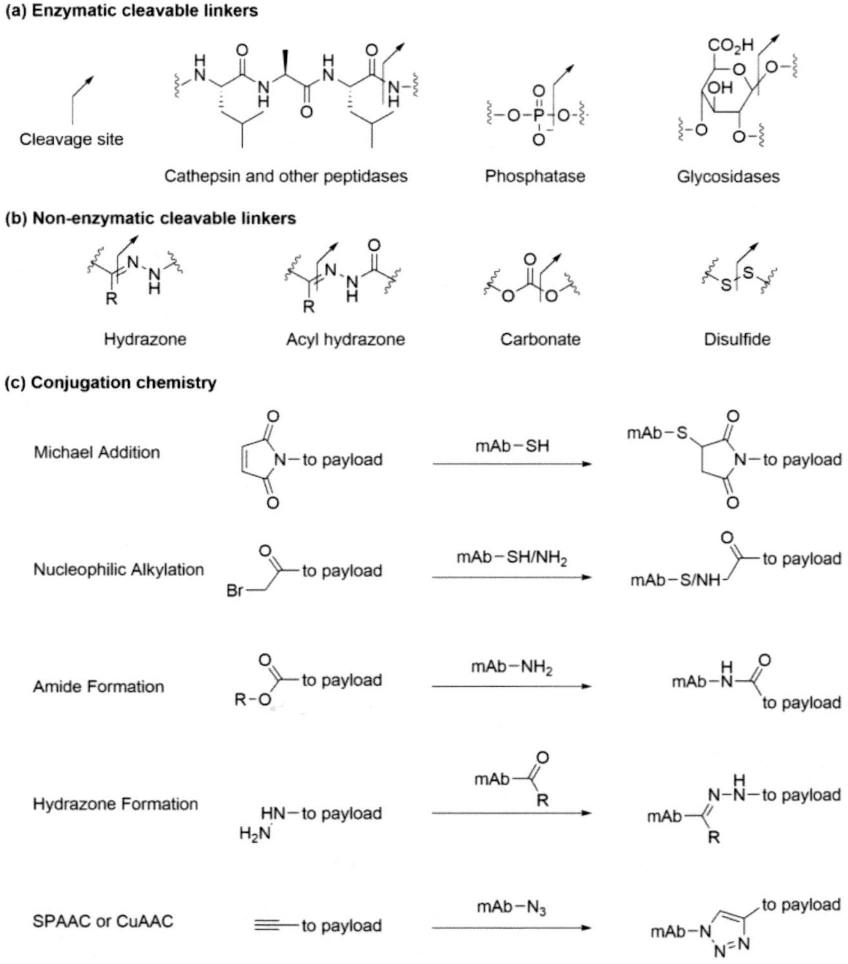

The use of unnatural amino acid *p*-acetylphenylalanine has been documented to form an oxime linkage with a linker with a hydroxylamine group. This oxime-containing ADC linker demonstrated lower toxicity and augmented stability and antitumor activity *in vivo* over the maleimide-conjugated ADC. Further antibody engineering has facilitated the use of new approaches such as copper-catalyzed azide alkyne cycloaddition (CuAAC) and strain promoted azide alkyne cycloaddition (SPAAC).

Enzymatic reactions can also be employed for attaching the antibody and liner-payload. For instance, glycosyltransferases, sortase, transglutaminase, farnesyltransferase and formylglycine generating enzyme. While the advantages are high efficiency and specificity, there are points that warrant further study such as substrate sequence in the antibody and longer time entailed.

Figure 10 below summarizes various conjugation systems (a) Random conjugation to primary amines, with hypothetical drug-to-antibody ratio (DAR) = 4. (b) Cysteine conjugation post-reduction of interchain disulfide bonds, with hypothetical DAR = 4. (c) Complete cysteine conjugation with DAR = 8 (e.g., trastuzumab deruxtecan). (d) Unnatural amino acid conjugation, DAR = 2. (e) Enzymatic conjugation to the introduced substrate sequence, DAR = 2: Shim H (2020) (© 2020 by the author. Licensee MDPI, Basel, Switzerland. *This article is an open access article distributed under the terms and conditions of the Creative Commons Attribution (CC BY) license: http://creativecommons.org/licenses/by/4.0/*):

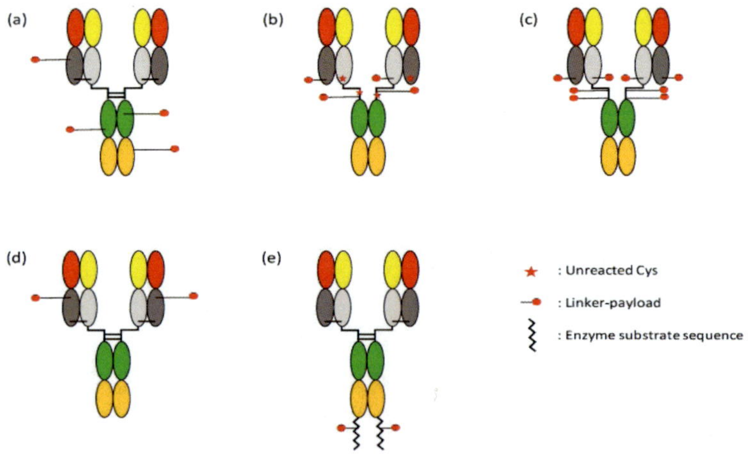

In order to diminish bystander killing, an ADC with diphosphatase-cleavable linker has been documented. This entailed the use of a payload of a fluticasone propionate derivative to immune cells documenting augmented *in vitro* potency. For hydrophobic payloads, linker design is being much explored. This entails boosting the hydrophilicity of the ADC by adding a polyethylene glycol (PEG) or a glucuronic acid unit or both (Leung D, Wurst JM, Liu T, Martinez RM, Datta-Mannan A, Feng Y, 2020).

Figure 11 below is illustrative of novel linker approaches; ***TOP***: Lower permeability (Leung D, Wurst JM, Liu T, Martinez RM, Datta-Mannan A, Feng Y, 2020: © 2020 by the authors. Licensee MDPI, Basel, Switzerland. *This article is an open access article distributed under the terms and conditions of the Creative Commons Attribution (CC BY) license: http:// creativecommons.org/licenses/by/4.0/*):

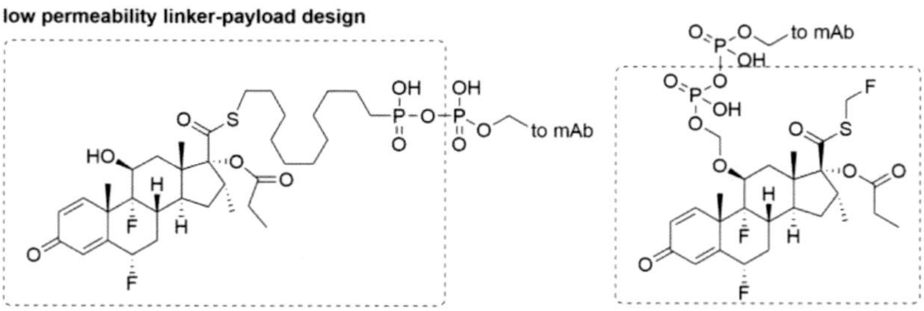

BOTTOM: a hydrophilic linker to tackle hydrophobicity (Leung D, Wurst JM, Liu T, Martinez RM, Datta-Mannan A, Feng Y, 2020: © 2020 by the authors. Licensee MDPI, Basel, Switzerland. *This article is an open access article distributed under the terms and conditions of the Creative Commons Attribution (CC BY) license: http://creativecommons.org/ licenses/by/4.0/*):

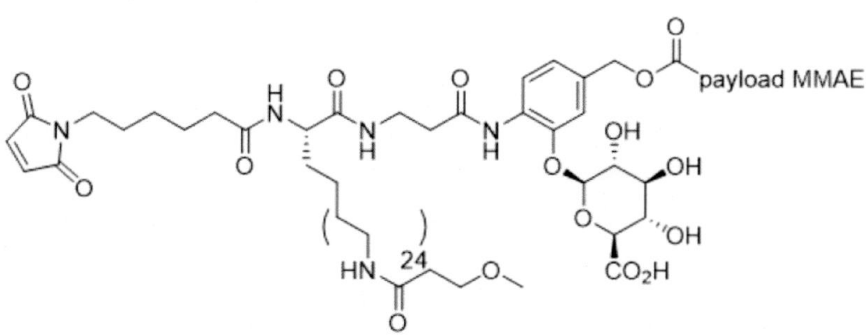

4.6 CASE STUDY OF A LINKER

2018-published research in *Nature Communications* by Anami and team reported the increased stability of a glutamic acid–valine–citrulline (EVCit) tripeptide sequence in mouse models over valine–citrulline (VCit) dipeptide linkers (the presence of extracellular carboxylesterase 1c (Ces1c) is attributed for this lowered stability of the latter in the murine system).

Increased stability of the ADCs designed was observed in human and mouse plasma while their cathepsin B-mediated cleavage sensitivity was maintained for intracellular payload release. Further *in vivo* probing in mouse models unveiled the therapeutic potential of ADCs albeit at lowered premature payload release for utility in preclinical screening.

Figure 12 below depicts the results of using a tripeptide linker in an ADC:

TOP: Plasma stability **a** in human plasma; **b** in mouse plasma; *in vitro* cytotoxicity in KPL-4 (**c**), JIMT-1 (**d**), and MDA-MB-231 (**e**) for unconjugated N297A anti-HER2 mAb (black cross), VCit ADC (**3a**, blue circle), SVCit ADC (**3b**, orange triangle), EVCit ADC (**3c**, green square), non-cleavable ADC (**4**, red inverted triangle), and isotype control ADC containing EVCit (**5**, magenta diamond, non-targeting control) (Anami et al (2018): *Open Access: This article is licensed under a Creative Commons Attribution 4.0 International License, which permits use, sharing, adaptation, distribution and reproduction in any medium or format, as long as you give appropriate credit to the original author(s) and the source, provide a link to the Creative Commons license, and indicate if changes were made. The images or other third party material in this article are included in the article's Creative Commons license, unless indicated otherwise in a credit line to the material. If material is not included in the article's Creative Commons license and your intended use is not permitted by statutory regulation or exceeds the permitted use, you will need to obtain permission directly from the copyright holder. To view a copy of this license, visit http://creativecommons.org/licenses/by/4.0/*):

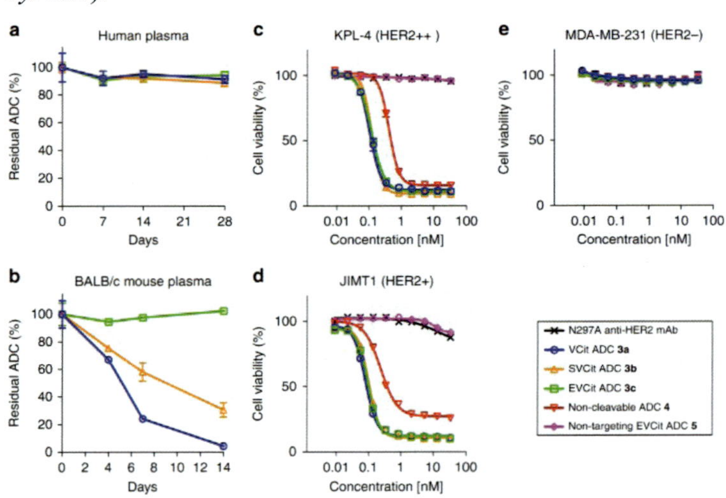

BOTTOM: *In vivo* pharmacokinetics (PK) of unconjugated N297A anti-HER2 mAb (black cross), VCit (blue circle), SVCit (orange triangle), and EVCit (green square) ADCs (**a**–**c**) in female BALB/c mice. Total antibody (conjugated and unconjugated, **a**) and ADC (conjugated only, **b**) by sandwich ELISA. **c**, **d** Antitumor activity of anti-HER2 ADCs (**3a**, **c**) in the JIMT-1 (**c**) and KPL-4 (**d**) xenograft tumor models. A single dose of VCit ADC (**3a**, 3 mg kg^{-1}, blue circle), EVCit ADC (**3c**, 3 mg kg^{-1}, green square; 1 mg kg^{-1}, magenta triangle), or vehicle control (gray inverted triangle). **e**, **f** Percentages of surviving mice over time in the JIMT-1 (**e**) and KPL-4 (**f**) models (Anami et al (2018): *Open Access: This article is licensed under a Creative Commons Attribution 4.0 International License, which permits use, sharing, adaptation, distribution and reproduction in any medium or format, as long as you give appropriate credit to the original author(s) and the source, provide a link to the Creative Commons license, and indicate if changes were made. The images or other third party material in this article are included in the article's Creative Commons license, unless indicated otherwise in a credit line to the material. If material is not included in the article's Creative Commons license and your intended use is not permitted by statutory regulation or exceeds the permitted use, you will need to obtain permission directly from the copyright holder. To view a copy of this license, visit http://creativecommons.org/licenses/by/4.0/*):

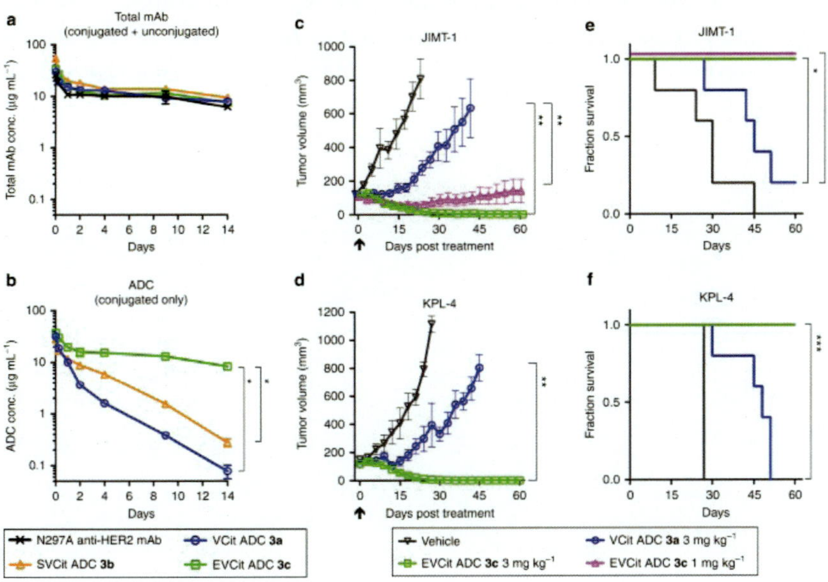

4.7 OLD AND NEW PAYLOADS

One can document the predominant use of auristatin E and emtansine (both are anti-mitotic) in FDA-approved ADCs like Kadcyla® and in preclinical assays as well as the DNA-damaging calicheamicin in Mylotarg®, duocarmycins, and PBDs. The payload is seeing promise in the form of a potent bis-intercalator depsipeptide (SW-163D: Pfizer), pyrrole-based KSP (kinesin spindle protein) inhibitors, topoisomerase I inhibitor (DXd derived from exatecan), and targeted delivery by nicotinamide phosphoribosyltransferase (NAMPT) inhibitors. Delivering a phosphodiesterase 4 (PDE4) inhibitor (GSK256066) via CD11 for specifically targeting leukemia with dasatinib and an anti-CXCR4 antibody is also reported. Given the use of glucocorticoids in immunology, auto-immune disorders also see promise in the form of specific antibodies linked to glucocorticoid receptor (GR) agonists.

Figure 13 below depicts new entrants as payloads in ADCs: SW-163D (a), pyrrole-based kinesin spindle protein (KSP) inhibitor (b), topoisomerase I inhibitor (DXd, c) NAMPT inhibitor (d), and MMP9 inhibitor (CGS27023A, e). For non-oncology payloads LXR agonist (f), PDE4 inhibitor (GSK256066, g), kinase inhibitor dasatinib (h), antimicrobial rifamycin analog (i), GR agonists dexamethasone (j), budesonide (k), and fluticasone propionate (i) ((Leung D, Wurst JM, Liu T, Martinez RM, Datta-Mannan A, Feng Y, 2020: © 2020 by the authors. Licensee MDPI, Basel, Switzerland. *This article is an open access article distributed under the terms and conditions of the Creative Commons Attribution (CC BY) license: http://creativecommons.org/licenses/by/4.0/*):

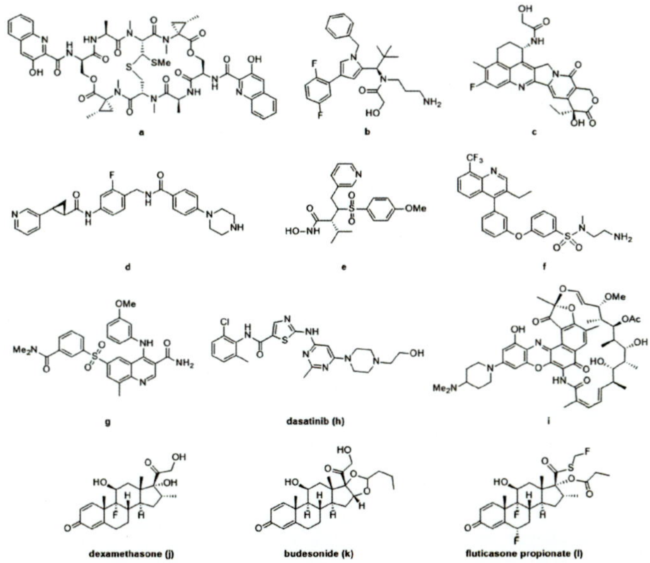

PBD dimers have been conjugated to cleavable linker drug called talirine; a maleimidocaproyl linker with a valine-alanine-PAB sensitive to cathepsin B (DAR=2). The absence of resistance against these molecules facilitates the use of these dimers for MMAE- or DM1-ADC-resistant malignancies. Figure 14 below depicts a pyrrolobenzodiazepine dimer (PBD; SGN-1882)-bearing ADC (Joubert N, Beck A, Dumontet C, Denevault-Sabourin C, 2020: © 2020 by the authors. Licensee MDPI, Basel, Switzerland. *This article is an open access article distributed under the terms and conditions of the Creative Commons Attribution (CC BY) license: http://creativecommons.org/licenses/by/4.0/*):

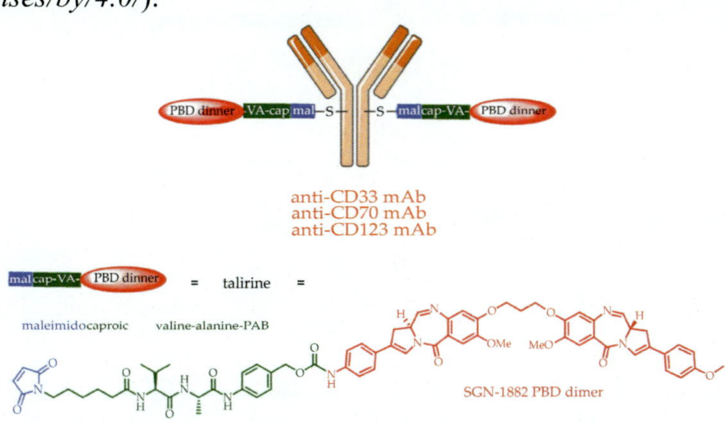

Tesirine as a payload has been explored with loncastuximab tesirine (ADCT-402) for against B cell acute lymphoblastic leukemia and camidanlumab tesirine (ADCT-301) for Hodgkin lymphoma in trials. Tesirine (SG3249) amalgamates optimal hydrophobicity and potent antitumor activity. Figure 15 below is illustrative of tesirine-based ADCs (Joubert N, Beck A, Dumontet C, Denevault-Sabourin C, 2020: © 2020 by the authors. Licensee MDPI, Basel, Switzerland. *This article is an open access article distributed under the terms and conditions of the Creative Commons Attribution (CC BY) license: http://creativecommons.org/licenses/by/4.0/*):

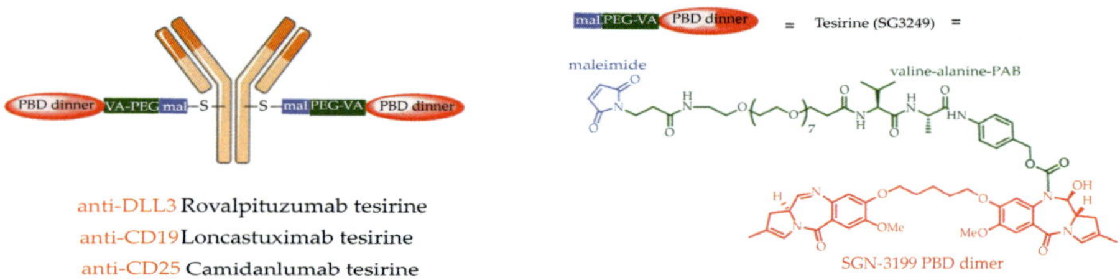

PNU-159682 from Genentech is a new anthracycline analog documented an IC_{50} of 20–100 pM. This was used in an anti-CD22-NMS249 ADC with the usual PAB linker and MC-VC-PAB-DEA (*N,N'*-dimethylethylenediamine) linker. Figure 16 below illustrates a PNU-159682-based ADC (Joubert N, Beck A, Dumontet C, Denevault-Sabourin C, 2020: © 2020 by the authors. Licensee MDPI, Basel, Switzerland. *This article is an open access article distributed under the terms and conditions of the Creative Commons Attribution (CC BY) license: http://creativecommons.org/licenses/by/4.0/*):

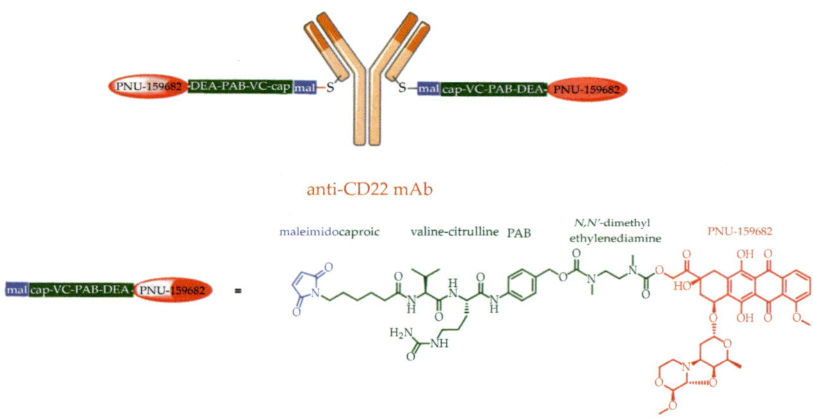

The advanced ATAC (antibody targeted amanitin conjugate) system based on amatoxins has seen the development of HDP-101 by Heidelberg

Pharma. This targets CD269 or the B-cell maturation antigen (BCMA) with a cathepsin B-sensitive linker and an α-amanitin analog (HDP 30.2115). Testing on patient melanoma cells revealed the observed IC_{50} of the ATAC HDP-101 at around 100 pM over the IC_{50} of 30.2115 for free amatoxin with an absence of toxicity on cells without BCMA paving the way for detailed elucidation.

Figure 17 below illustrates HDP-101(Joubert N, Beck A, Dumontet C, Denevault-Sabourin C, 2020: © 2020 by the authors. Licensee MDPI, Basel, Switzerland. *This article is an open access article distributed under the terms and conditions of the Creative Commons Attribution (CC BY) license: http://creativecommons.org/licenses/by/4.0/*):

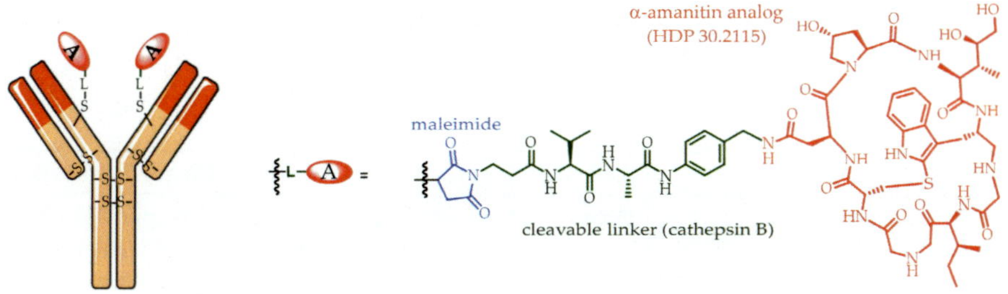

anti-BCMA **THIOMAB-amanitin (HDP-101)**

4.8 IMPORTANCE OF ENDOCYTOSIS

Probing and comprehending the endocytosis of target receptors and the internalization of ADCs becomes pertinent for ADCs given that the efficiency of this uptake and subsequent lysosomal trafficking are also vital aspects like the pharmacokinetics. Given that research is seeing a slew of ADCs in trials, augmenting the therapeutic index necessitates ADC internalization apart from drug, dosage and effects.

Currently approved ADCs entail several endocytosis pathways as outlined by figure 18 below and table 5 below Narrow arrows depict little utilization by HER2. Dashed arrows are illustrative of indirect evidence. * CD 33 can be independent of AP-2 despite the use of clathrin-mediated endocytosis (CME). Green bars are CD30 Hammood M, Craig AW, Leyton JV (2021) (© 2021 by the authors. Licensee MDPI, Basel, Switzerland. *This article is an open access article distributed under the terms and conditions of the Creative Commons Attribution (CC BY) license: https://creativecommons.org/licenses/by/4.0/*):

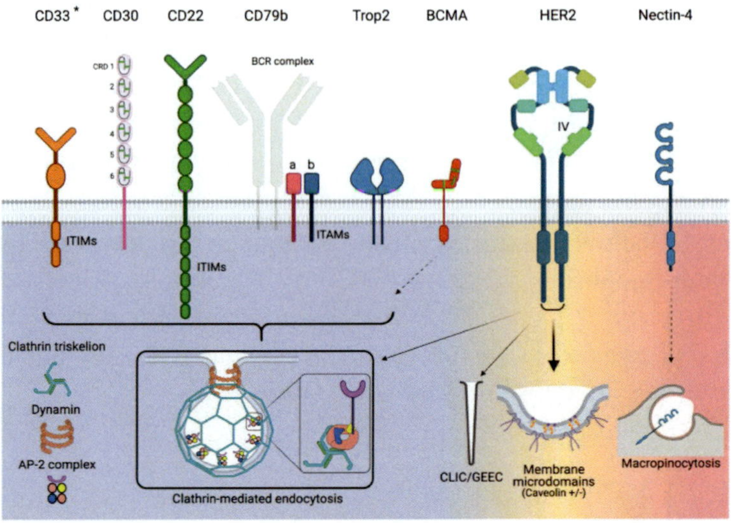

Receptor	Pathway	Activity	Association with ADC Efficacy/Resistance
CD33	CME	Poor	• AML patients who do not respond to GO have been linked to poor receptor internalization
CD30	CME	Poor	• Undergoes significant shedding from cell surface
CD22	CME	Good	• Fast endocytosis activates intracellular pools, which replenish the level of CD22 expression
CD79b	CME	Good	• Due to rapid internalization and trafficking to lysosomes, patients will most likely respond to PV treatment
Trop2	CME	Good	• Strong preclinical data link internalization to efficacy
BCMA	Insufficient information	Good	• Insufficient information
HER2	Clathrin-independent (caveolae +/−)	Poor	• Poor internalization linked with poor clinical outcomes • Dysregulation of the endocytotic machinery has been linked to resistance in preclinical models • Novel strategies such as induced HER2 crosslinking to improve endocytosis are currently in clinical testing
Nectin-4	Macropinocytosis	Good	• Insufficient information

4.9 NEW SYSTEMS

Given that the non-necessity Fc part of an IgG is being pursued for ADC construction by some research groups, this brings in smaller constructs such as single chain variable fragments (scFv), antigen binding fragments (Fab), peptides and scFv dimerized or small immunoproteins (SIP). Novel scFv-drug conjugates (SDCs) were developed from an anti-HER2 scFv (derived from trastuzumab) to an auristatin derivative. The initial promise of these SDCs is demonstrated by their ability to target HER2 (EC_{50} of 0.68 nM) vs. native scFv (0.32 nM) in *in vitro* SK-BR-3 HER2-positive cells.

Figure 19 below depicts an SDC (DAR=1) (Joubert N, Beck A, Dumontet C, Denevault-Sabourin C, 2020: © 2020 by the authors. Licensee MDPI, Basel, Switzerland. *This article is an open access article distributed under the terms and conditions of the Creative Commons Attribution (CC BY) license: http://creativecommons.org/licenses/by/4.0/*):

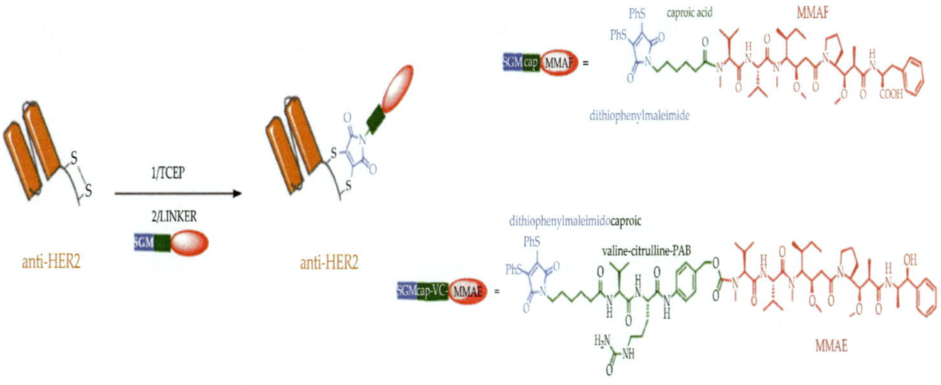

MedImmune has developed the biparatopic ADC IMMU4276 to address the intratumoral heterogeneity of HER2 causing relapse. A tubulysine derived powerful cytotoxic agent AZ13599185 with a maleimidocaproyl linker with a bivalent biparatopic mAb to target trastuzumab and pertuzumab-recognized epitopes. Figure 20 below depicts this biparatopic ADC (Joubert N, Beck A, Dumontet C, Denevault-Sabourin C, 2020: © 2020 by the authors. Licensee MDPI, Basel, Switzerland. *This article is an open access article distributed under the terms and conditions of the Creative Commons Attribution (CC BY) license: http://creativecommons.org/licenses/by/4.0/*):

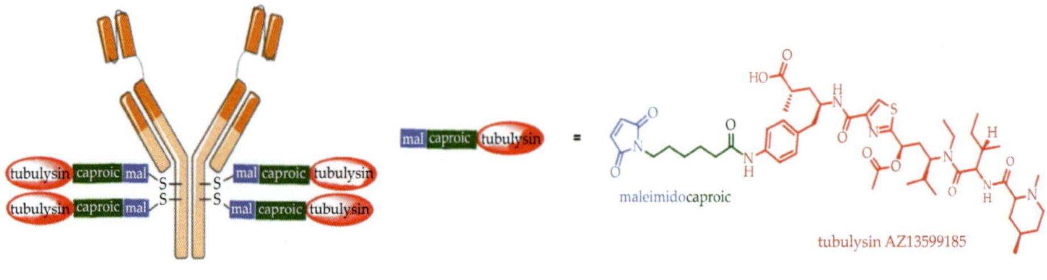

anti-HER2

An ADC (DAR=2) was developed using deglycosylation of an asparagine residue followed by transglutaminase-based addition to the mAb and an amino and nitrogen heterobifunctional linker. Copper-free Huisgen cycloaddition (SPAAC) facilitates the addition of the azide conjugate with a linker bearing dibenzylcyclooctyne (DBCO) and MMAE. This system is illustrated in figure 21 below (Joubert N, Beck A, Dumontet C, Denevault-Sabourin C, 2020: © 2020 by the authors. Licensee MDPI, Basel, Switzerland. *This article is an open access article distributed under the terms and conditions of the Creative Commons Attribution (CC BY) license: http://creativecommons.org/licenses/by/4.0/*):

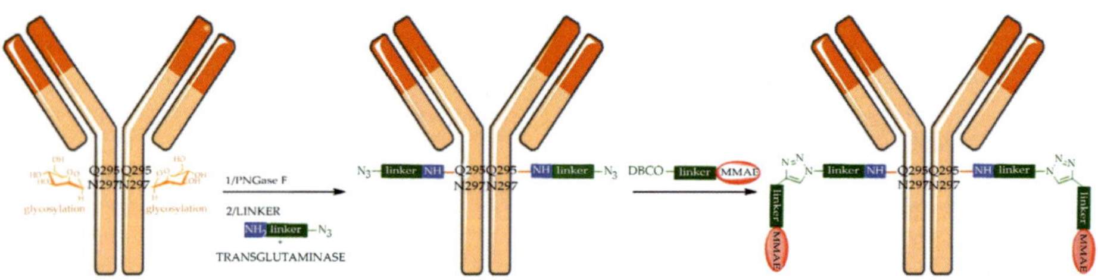

anti-CD30 mAb anti-CD30 mAb anti-CD30 mAb

In another approach, the homogeneity and stability of ADCs can be augmented by regiospecific bioconjugation. This entails the use of dibromomaleimide (DBM) or dithiophenylmaleimide (DSPh) linkers. The improvement in the therapeutic index and pharmacokinetic properties by DBM over the conventional maleimide linker system has been documented. This is shown in figure 22 below (Joubert N, Beck A, Dumontet C, Denevault-Sabourin C, 2020: © 2020 by the authors. Licensee MDPI, Basel, Switzerland. *This article is an open access article distributed under the terms and conditions of the Creative Commons Attribution (CC BY) license: http://creativecommons.org/licenses/by/4.0/*):

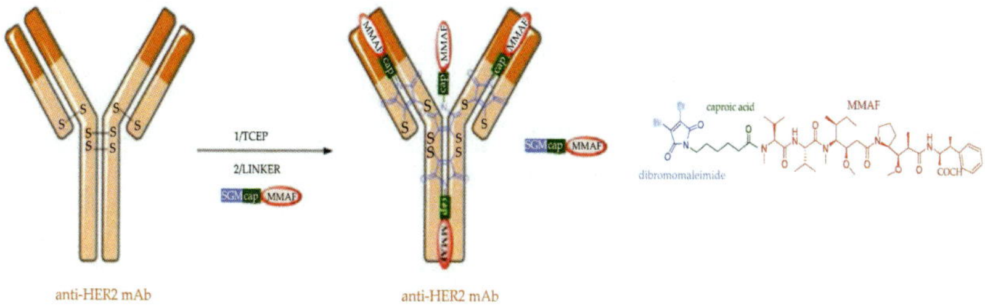

4.10 NANOTECHNOLOGY

A slew of advantages are documented for nanoparticles (NPs) ranging from selective targeting on account of high surface/volume ratio and controlled release of the drug. Specific antibodies can be conjugated to generate vectorized NPs that undergo internalization and release of the drug inside the target cell. This leads to the development of antibody-conjugated nanoparticles (ACNPs).

The overview of ACNPs is depicted in figure 23 below (Juan, A., Cimas, F. J., Bravo, I., Pandiella, A., Ocaña, A., & Alonso-Moreno, C. (2020): © 2020 by the authors. Licensee MDPI, Basel, Switzerland. *This article is an open access article distributed under the terms and conditions of the Creative Commons Attribution (CC BY) license: http://creativecommons.org/licenses/by/4.0/*):

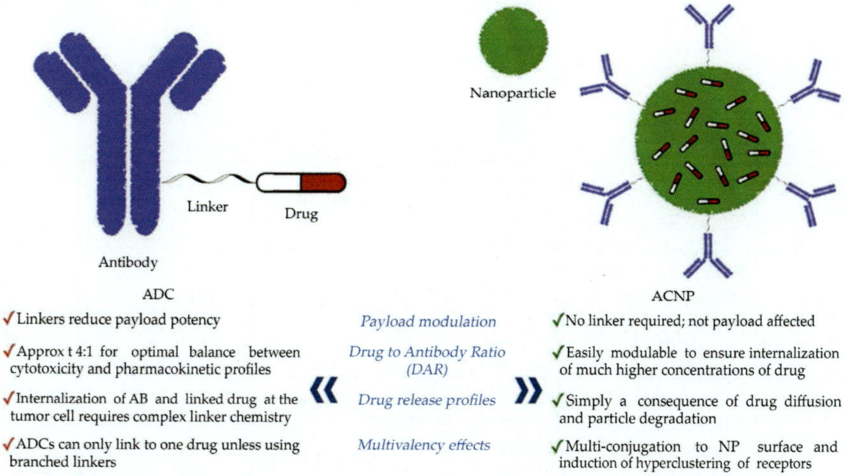

Both targeted therapies (ADCs and ACNPs) documenting mechanisms on the same lines as shown below in figure 24 (Juan, A., Cimas, F. J., Bravo,

I., Pandiella, A., Ocaña, A., & Alonso-Moreno, C. (2020): © 2020 by the authors. Licensee MDPI, Basel, Switzerland. *This article is an open access article distributed under the terms and conditions of the Creative Commons Attribution (CC BY) license: http://creativecommons.org/licenses/by/4.0/*):

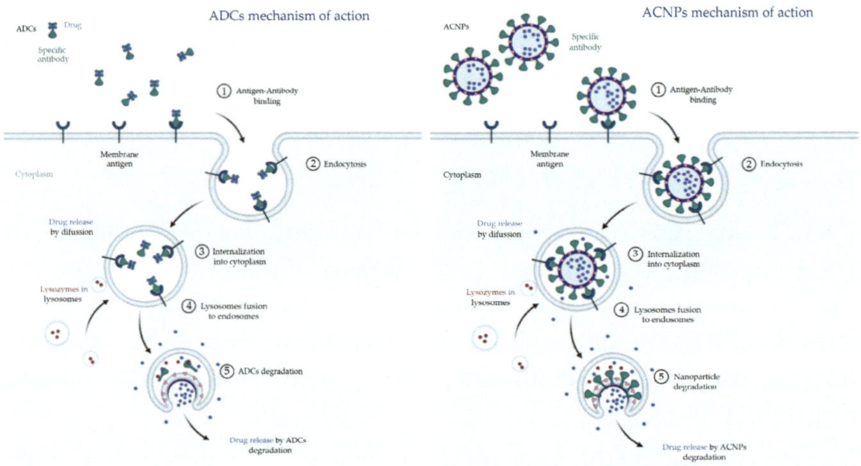

For targeting breast cancer, the main approaches of targeted therapy are summarized in table 6 below (Juan, A., Cimas, F. J., Bravo, I., Pandiella, A., Ocaña, A., & Alonso-Moreno, C, 2020: © 2020 by the authors. Licensee MDPI, Basel, Switzerland. *This article is an open access article distributed under the terms and conditions of the Creative Commons Attribution (CC BY) license: http://creativecommons.org/licenses/by/4.0/*):

Conjugation Method	**Nanoparticle Characteristic**	**Drug Encapsulated**	**Antibody Attached (Cross-Linker and or Substrategy)**	**Reference**
Carbodiimide Chemistry	Magnetic poly(d,l-lactide-*co*-glycolide)	Doxorubicin	Trastuzumab	[59]
	Poly(lactic-*co*-glycolic acid) (PLGA)		Anti MCF-7 mAb	[67]
	Poly(d,l-lactide-*co*-glycolide) (PLGA)	Rapamycin	monoclonal EGFR antibody	[69]
	Poly(lactide)-d-α-tocopheryl polyethylene glycol succinate (PLA-TPGS/TPGS–COOH) copolymers	Docetaxel and iron oxides	Trastuzumab	[60]

	Poly(lactide)-d-α-tocopheryl polyethylene glycol succinate (PLA-TPGS/TPGS-COOH) copolymers	Docetaxel	Trastuzumab	[61]
	Amine terminated d-α-tocopheryl polyethylene glycol 1000 succinate (TPGS–NH_2)	Docetaxel	Cetuximab	[69]
	Poly(N-vinylpyrrolidone/ Poly(d,l-lactic-*co*-glycolic acid) (PVP–PLGA)	Tamoxifen	Trastuzumab	[62]
	Folate conjugated poly(d,l-lactide) (PLA) polyethylene glycol (PEG) PLA-PEG-PLA-urthane-s-s random multiblock copolymer	Doxorubicin	Trastuzumab	[63]
	(PEG-β-PLGA) copolymer	Indocyanine green and doxorubicin	Anti-HER2	[70]
	Poly(lactic-*co*-glycolic acid) (PLGA)	Epirubicin	Trastuzumab	[64]
	Poly(ethylene glycol)-poly(ε-caprolactone) copolymer (PEG-PCL)	Dasatinib	Trastuzumab	[65]
	Poly(lactic-*co*-glycolic acid) (PLGA)	Paclitaxel	Trastuzumab	[66]
Maleimide Chemistry	Human Serum Albumin (HSA)		Trastuzumab (NHS-PEG5000-Mal)	[154]
	Poly(2-methyl-2-carboxytrimethylene carbonate-*co*-d,l-lactide)-graft-poly(ethylene glycol)-furan (Poly(TMCC-*co*-LA)-*g*-PEG-furan)	Doxorubicin	Trastuzumab	[118]
	Poly(d,l-lactide-*co*-glycolide) (PLGA)	Docetaxel	Trastuzumab (NHS-PEG-Mal)	[155]
	Chitosan	Doxorubicin	Trastuzumab (sulfosuccinimidyl 4-(N-maleimidomethyl) cyclohexane-1-1-carboxylate)	[86]
	Bovine serum albumin (BSA)	5-fluorouracil	PR81 (NHS-PEG7500-Mal)	[156]
	Poly(d,l-lactide-*co*-glycolide) (PLGA)	Paclitaxel	anti-CD133 (NHS-PEG-Mal)	[157]
	Poly(d,l-lactide-*co*-glycolide) (PLGA)	Paclitaxel	Trastuzumab (DSPE-PEG2000-Mal)	[88]
	Poly(d,l-lactide-*co*-glycolide) (PLGA)	Paclitaxel	Clone 6, AM6 (NHS-PEG-Mal)	[158]

	Polyethyelenimine-polyethylen-glycol copolymer (PEI-PEG)	SiRNA	anti-HER2 Nb (RR4) (NHS-PEG3500-Mal)	[159]
Click Chemistry	(Poly(TMCC-*co*-LA)-*g*-PEG-furan)		Trastuzumab (Diels-Alder)	[117]
	(Poly(TMCC-*co*-LA)-*g*-PEG-furan)	Doxorubicin	Trastuzumab (Diels-Alder)	[118]
	(PLGA-β-PEG-azide and PLGA-β-PHis-β-PEG-azide)	Doxorubicin	Trastuzumab (CuAAC)	[111]
	Poly(d,l-lactide-*co*-2-methyl-2-carboxytrimethylene carbonate) (P(LA-*co*-TMCC)	Docetaxel	Fab 73J (Diels-Alder)	[109]
	(PLGA–PEG-azide)		Trastuzumab (2,5-Dioxo-pyrrolidin-1-yl 1-((1R,8S,9S)-bicyclo[6.1.0] non-4-yn-9-yl)-3,14-dioxo-2,7,10-trioxa-4,13-diazaoctadecan-18-oate/SPAAC)	[89]
	Poly(d,l-lactide-coglycolide)-β-polyethylene glycol (PLGA-PEG)		Trastuzumab (CuAAC)	[112]
Aptamers	Human serum albumin (HSA)		Trastuzumab (Biotin/Neutravidin)	[142]
	Methoxy poly(ethylene glycol)-b-poly(ε-caprolactone) and methoxy poly(ethylene glycol)-b-poly(ε-caprolactone)	Doxorubicin and quercetin		[116]
	(PLGA-PEG-biotin copolymer)	SN-38		[117]
	Poly(d,l-lactide-coglycolide) (PLGA) and Poly(d,l-lactide-coglycolide) (PLGA)-β-polyethylene glycol (PEG)	Doxorubicin and p-gp SiRNA	Aptamer A6 (DSPE-PEG-Mal)	[143]
	Avidin-Nucleic-Acid-Nano-Assemblies (ANANAS)	Biotin-PEG-Atto488, biotin- Hz-doxo and biotin-PEG-Hz-doxo	Biotin-PEG-cetuximab, Cetuximab-Atto488 and cetuximab-Hz-doxorubicin	[13

4.11 FINAL POINTS

ADCs are receiving much interest from scientists. To give an example, table 7 below lists ongoing clinical trials probing ADCs for breast cancer treatment (Juan, A., Cimas, F. J., Bravo, I., Pandiella, A., Ocaña, A., & Alonso-Moreno, C, 2020: © 2020 by the authors. Licensee MDPI, Basel, Switzerland. *This article is an open access article distributed under the terms and conditions of the Creative Commons Attribution (CC BY) license: http://creativecommons.org/licenses/by/4.0/*):

Breast Cancer Subtype	**Clinical Trial**	**Agent**	**Target**	**Combinatorial Agent**	**Phase**
Luminal	NCT03874325	Durvalumab	PD-L1	Anastrozole	II
	NCT03409198	Chemo + ipilim-umab + nivolumab	CTLA-4/PD-1	Ipilimumab, nivolumab, pegylated liposomal doxorubicin and cyclophosphamide	II
	NCT03691311	Denosumab	RANKL	-	I
	NCT03879174	Pembrolizumab	PD-1	Tamoxifen	II
	NCT03051659	Pembrolizumab	PD-1	Eribulin Mesylate	II
	NCT03393845	Pembrolizumab	PD-1	Fulvestrant	II
	NCT03492918	Pembrolizumab	PD-1	-	II
	NCT03608865	Durvalumab and Tremelimumab	PD-L1 and CTLA-4	-	II
	NCT04251169	Pembrolizumab	PD-1	Paclitaxel	II
	NCT01491737	Pertuzumab and Trastuzumab	HER2	Anastrozole	II
	NCT04088032	Durvalumab	PD-L1	Abemaciclib and Anastrozole	I
	NCT03241810	Seribantumab	HER3	Fulvestrant	II
	NCT03132467	Durvalumab and Tremelimumab	PD-L1 and CTLA-4	-	I
	NCT00405938	Bevacizumab	VEGF-A	Anastrozole	II
	NCT03659136	Xentuzumab	IGF	Everolimus	II
	NCT02990845	Pembrolizumab	PD-1	Exemestane/Leup-rolide	I/II
	NCT00022672	Trastuzumab	HER2	anastrazole	III
	NCT03099174	Xentuzumab	IGF	Letrozole/anastrozole	I
	NCT02971748	Pembrolizumab	PD-1	-	II
	NCT03051672	Pembrolizumab	PD-1	Palliative radiotherapy	II
	NCT02997995	Durvalumab	PD-L1	-	II
	NCT04243616	Cemiplimab	PD-1	Cemiplimab + Pacli-taxel + Doxorubicin + Cyclophosphamide	II
	NCT04243616	U3-1402	HER3	-	I/II
HER2	NCT03321981	MCI-A-128	HER2	Trastuzumab/Chemo-therapy	II

	NCT03983395	GBR 1302	CD3/HER2	-	I/II
	NCT03052634	RC48	HER2	-	Ib/II
	NCT03523585	DS-8201a	HER2	-	III
	NCT03529110	DS-8201a	HER2	-	III
	NCT03734029	DS-8201a	HER2	-	III
	NCT03052634	RC48	HER2	-	Ib/II
	NCT03944499	FS-1502	HER2	-	I
	NCT03262935	SYD985	HER2	-	II
	NCT03255070	ARX788	HER2	-	I
	NCT03032107	Pembrolizumab	PD-1	T-DM1	Ib/II
	NCT03747120	Pembrolizuab	PD-1	Paclitaxel, Trastuzumab and Pertuzumab	II
	NCT02605915	Atezolizumab	PD-1	Pertuzumab, trastumuzab	Ib
	NCT03199885	Atezolizumab	PD-1	Paclitaxel, Trastuzumab and Pertuzumab	III
	NCT03125928	Atezolizumab	PD-1	Paclitaxel, Trastuzumab and Pertuzumab	IIa
	NCT03726879	Atezolizumab	PD-1	Doxorubicin, Cyclophosphamide, Paclitaxel, Trastuzumab and Pertuzumab	III
	NCT03417544	Atezolizumab	PD-1	Pertuzumab, trastumuzab	II
	NCT04034823	KN035	PD-1	Trastuzumab/ Docetaxel	II
	NCT03112590	IFN-Y	IFN-Y	Paclitaxel, Trastuzumab and Pertuzumab	II
	NCT03135171	Tocilizumab	IL-6	Trastuzumab and Pertuzumab	I
	NCT03364348	Utomilumab	TNF	Trastuzumab emtansine or Trastuzumab	Ib
	NCT03414658	Utomilumab	TNF	Vinorelbine, Avelumab, Utomilumab	II
TNBC	NCT02655822	CPI-444	A2AR	Atezolizumab (PD-LI)	I/Ib
	NCT03454451	CPI-006	CD73	Pembrolizumab CD-I), cm-444 (A2AR)	I
	NCT03251313	JS001	PD-1	Gemcitabine and cisplatin	I
	NCT03012230	Pembrolizumab	PD-1	Ruxolitinib OAK)	I
	NCT02890069	Spartlizumab	PD-1	Everolimus (mTOR), panobinosat (HDAC), LCL161 (apoptosis), QBM076 (CXCR2)	I
	NCT03250832	TSR-033	LAG-3	Anti-PD-1 antibodies	I

	NCT02646748	Pembrolizumab	PD-1	Itacitinib OAK), INCB050465 (P13K)	I
	NCT02947165	NIS793	TGFß	Spartalizumab (PD-1)	I/Ib
	NCT03549000	NZV930	CD73	Spartalizumab (PD-1), NIR178 (A2AR)	I
	NCT02838823	JS001	PD-1	-	I
	NCT02622074	Pembrolizumab	PD-1	nab-Paclitaxel, doxo-rubicin,	I
	NCT03292172	Atezolizumab	PD-L1	cyclophospha-mide, carboplatin R06870810 (BET)	I
	NCT02936102	FAZ053	PD-L1	Spartalizumab (PD-I)	I
	NCT03579472	M7824	PD-Ll/TGFß	Eribulin	I
	NCT0280744	MCSI 10	CSF-I	Spartalizumab (PD-1)	I
	NCT024602.24	LAG525	LAG-3	Spartalizumab (PD-1)	I/II
	NCT03241173	INCAGN01949	OX-40	Nivolumab (anti-PD-1) and/or ipilimumab (anti-CTLA4)	I/II
	NCT035912.76	Pembrolizumab	PD-1	Pegylated doxorubicin	I/II
	NCT02628132	Durvalumab	PD-L1	Paclitaxel	I/II
	NCT02657889	Pembrolizumab	PD-1	Niraparib (PARP)	I/II
	NCT03356860	Durvalumab	PD-L1	Paclitaxel, epirubicin, cyclophosphamide	I/II
	NCT02513472	Pembrolizumab	PD-1	Eribulin	I/II
	NCT02484404	Durvalumab	PD-L1	Olaparib (PARP)	I/II
	NCT02708680	Atezolizumab	PD-L1	Entinostat (HDAC)	I/II
	NCT02734004	Durvalumab	PD-L1	Diaparib (PARD	I/II
	NCT02614833	IMP321	LAG-3	Paclitaxel	II
	NCT03394287	SHR-1210	PD-1	Apatanib (VEGFR)	II
	NCT03414684	Nivolumab	PD. 1	Carboplatin	II
	NCT02648477	Pembrolizumab	PD-1	Doxorubicin	II
	NCT03004183	Pembrolizumab	PD-1	SBRT and ADV/ HSV-tk	II
	NCT02536794	Durvalumab	PD-LI	Tremelimumab (CTLA-4)	II
	NCT02752685	Pembrolizumab	PD-1	nab-Paclitaxel	II
	NCT03095352	Pembrolizumab	PD-1	Carboplatin	II
	NCT03184558	Pembrolizumab	PD-1	Bemcentinib (AXL)	II
	NCT02971761	Pembrolizumab	PD-1	Enobosarrn (selecdve androgen receptor modulator)	II
	NCT02554812	Avelumab	PD-L1	Utomilumab (CD 137)	II
	NCT02849496	Atezolizumab	PD-L1	Olaparib (PARP)	II

	NCT03483012	Atezolizumab	PD-L1	Stereotactic radiosurgery	II
	NCT03164993	Atezolizumab	PD-L1	Pegytated doxorubicin	II
	NCT01898117	Atezolizumab	PD-L1	Carboplatin and cyclophosphamide or paclitaxel	II
	NCT02883062	Atezolizumab	PD-L1	Carboplatin	II
	NCT02981303	Pembrolizumab	PD-1	Imprime PGG (PAMP)	II
	NCT02819518	Pembrolizumab	PD-1	nab-Paclltaxel, gemcitabine, carboplatin	III
	NCT03498716	Atezolizumab	PD-L1	Paclitaxel, epirubicin. cyclophosphamide	III
	NCT03197935	Atezolizumab	PD-L1	nab-Paclitaxel, doxorubicin,	III
	NCT02425891	Atezolizumab	PD-L1	cyclophosphamide nab-Paclitaxel	III
	NCT03125902	Atezolizumab	PD-L1	Paclitaxel	III
	NCT02574455	Sacituzumab govitecan	Trop-2	-	III

ADCs have entered the clinical field for treatment and studies on these "bullets" are ongoing. The detailed scrutiny of linkers, payloads and conjugation approaches can further augment the development of more ADCs to enter the market to quell not only cancer but also other diseases.

"Research is seeing what everybody else has seen and thinking what nobody else has thought."

- Albert Szent-Györgyi

4.12 REFERENCES

1. Hammood M, Craig AW, Leyton JV. Impact of Endocytosis Mechanisms for the Receptors Targeted by the Currently Approved Antibody-Drug Conjugates (ADCs)—A Necessity for Future ADC Research and Development. *Pharmaceuticals*. 2021; 14(7):674. https://doi.org/10.3390/ph14070674
2. Joubert N, Beck A, Dumontet C, Denevault-Sabourin C. Antibody–Drug Conjugates: The Last Decade. *Pharmaceuticals*. 2020; 13(9):245. https://doi.org/10.3390/ph13090245
3. Tang H, Liu Y, Yu Z, Sun M, Lin L, Liu W, Han Q, Wei M and Jin Y (2019) The Analysis of Key Factors Related to ADCs Structural Design. *Front. Pharmacol.* 10:373. doi: 10.3389/fphar.2019.00373
4. Hafeez, U., Parakh, S., Gan, H. K., & Scott, A. M. (2020). Antibody-Drug Conjugates for Cancer Therapy. *Molecules (Basel, Switzerland)*, *25*(20), 4764. https://doi.org/10.3390/molecules25204764
5. Collins, D. M., Bossenmaier, B., Kollmorgen, G., & Niederfellner, G. (2019). Acquired Resistance to Antibody-Drug Conjugates. *Cancers*, *11*(3), 394. https://doi.org/10.3390/cancers11030394
6. Ward E.S., Ober R., Kang J., Sun W. Endolysosomal targeting conjugates for improved delivery of cargo molecules to the endolysosomal compartment of target cells. *Patent WO2018136455A1*. :2018.
7. Iwata TN, Sugihara K, Wada T, Agatsuma T (2019) [Fam-] trastuzumab deruxtecan (DS-8201a)-induced antitumor immunity is facilitated by the anti–CTLA-4 antibody in a mouse model. PLoS ONE 14(10): e0222280. https://doi.org/10.1371/journal.pone.0222280
8. Yamazaki, C.M., Yamaguchi, A., Anami, Y. *et al.* Antibody-drug conjugates with dual payloads for combating breast tumor heterogeneity and drug resistance. *Nat Commun* 12, 3528 (2021). https://doi.org/10.1038/s41467-021-23793-7
9. Shim H. (2020). Bispecific Antibodies and Antibody-Drug Conjugates for Cancer Therapy: Technological Considerations. *Biomolecules*, *10*(3), 360. https://doi.org/10.3390/biom10030360
10. Leung D, Wurst JM, Liu T, Martinez RM, Datta-Mannan A, Feng Y. Antibody Conjugates-Recent Advances and Future Innovations. *Antibodies*. 2020; 9(1):2. https://doi.org/10.3390/antib9010002
11. Anami, Y., Yamazaki, C.M., Xiong, W. *et al.* Glutamic acid–valine–citrulline linkers ensure stability and efficacy of antibody–

drug conjugates in mice. *Nat Commun* 9, 2512 (2018). https://doi.org/10.1038/s41467-018-04982-3

12. Juan, A., Cimas, F. J., Bravo, I., Pandiella, A., Ocaña, A., & Alonso-Moreno, C. (2020). Antibody Conjugation of Nanoparticles as Therapeutics for Breast Cancer Treatment. *International journal of molecular sciences*, *21*(17), 6018. https://doi.org/10.3390/ijms21176018
13. Juan, A., Cimas, F. J., Bravo, I., Pandiella, A., Ocaña, A., & Alonso-Moreno, C. (2020). An Overview of Antibody Conjugated Polymeric Nanoparticles for Breast Cancer Therapy. *Pharmaceutics*, *12*(9), 802. https://doi.org/10.3390/pharmaceutics12090802

CHAPTER 5

A Snapshot of other Biotherapeutics

"Evolution lies at the heart of biology. It is seamlessly and continuously linked to health research to better understand such conditions as AIDS or bird flu or Parkinson's or cancer or heart disease. Every biomedical experiment, every tiny advance, every major breakthrough ultimately connects to the principles first postulated by Darwin"

–Huntington Willard

Contents

5.1 CYTOKINES

Serving as factors for growth, differentiation, inflammation and anti-inflammation, are the polypeptides or glycoproteins with a molecular weight of below 30 kDa are "cytokines". While some of them are synthesized continuously like hematopoietic growth factors, they mostly are synthesized as a response to a particular stimulus to exert "short-term" actions. The presence of high-affinity receptors on the target cells facilitates the binding of cytokines to set off signaling circuits to alter gene transcription. Based on the cytokine(s) levels and exposure, the response of the receptors is varied to hence document antagonism or synergy among different cytokines (Berraondo et al, 2019).

Table 1 below documents approved cytokines for humans (Frazzini, S., Riva, F., & Amadori, M. (2021): © 2021 by the authors. Licensee MDPI, Basel, Switzerland. *This article is an open access article distributed under the terms and conditions of the Creative Commons Attribution (CC BY) license: https://creativecommons.org/licenses/by/4.0/*):

Key: IFN = interferon; TNF-α = tumor necrosis factor-alpha; G-CSF = granulocyte colony-stimulating factor; GM-CSF = granulocyte-macrophage colony-stimulating factor; IL = interleukin; EPO = erythropoietin

Cytokines	Target	References
IFN-α	Hepatitis B; hepatitis C	[35,36]
IFN-β	Multiple sclerosis	[37]
IFN-γ	Chronic granulomatous disease; Crohn's disease	[38,39]
TNF-α	Rheumatoid arthritis, psoriasis; Crohn's disease; ankylosing spondylitis; chronic obstructive pulmonary disease; sepsis; juvenile idiopathic arthritis; asthma	[40,41,42,43]
G-CSF	Febrile neutropenia; bone marrow transplantation	[44]
GM-CSF	Neutropenia after chemotherapy	[45]
IL-1	Rheumatoid arthritis; juvenile idiopathic arthritis; Still's disease; Crohn's disease	[42,46]
IL-2	Metastatic renal cell carcinoma; renal transplantation	[47,48,49]
IL-4	Asthma	[50]
IL-5	Asthma	[51]
IL-6	Rheumatoid arthritis	[52,53]
IL-8	Melanoma	[54]
IL-10	Crohn's disease; rheumatoid arthritis; psoriasis; ulcerative colitis; multiple sclerosis	[55]
IL-11	Thrombocytopenia; ulcerative colitis; psoriasis; rheumatoid arthritis; Crohn's disease	[56]
EPO	Anemia	[57]

In order to target malignancies, comprehensive elucidation of the immune system is the need of the hour on account of the low immunogenicity of tumors coupled with various aspects of the tumor microenvironment that impact immune cell functioning. The selection of systems to target malignancies is spawned by several such "immune modulation".

A slew of markers in immunology in the context of malignancies is represented below in figure 1 (© 2020 Chulpanova, Kitaeva, Green, Rizvanov and Solovyeva. *This is an open-access article distributed under the terms of the Creative Commons Attribution License (CC BY). The use, distribution or reproduction in other forums is permitted, provided the original author(s) and the copyright owner(s) are credited and that the original publication in this journal is cited, in accordance with accepted academic practice. No use, distribution or reproduction is permitted which does not comply with these terms*):

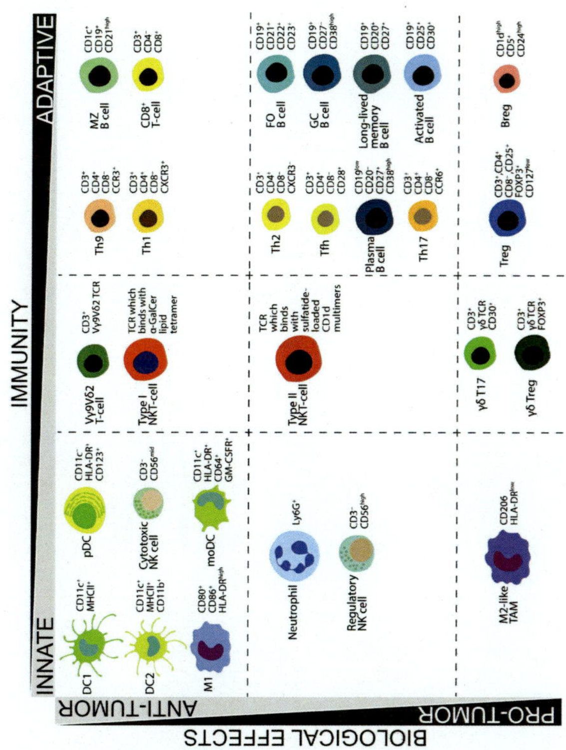

The impact of cytokines on various immune cells as unveiled by research is illustrated in figure 2 below (© 2020 Chulpanova, Kitaeva, Green, Rizvanov and Solovyeva. *This is an open-access article distributed under the terms of the Creative Commons Attribution License (CC BY). The*

use, distribution or reproduction in other forums is permitted, provided the original author(s) and the copyright owner(s) are credited and that the original publication in this journal is cited, in accordance with accepted academic practice. No use, distribution or reproduction is permitted which does not comply with these terms):

Immune cells	Cytokines						
	IL2	IL12	IL15	IL21	Type I IFNs	IFN-γ	GM-CSF
Innate immunity cells: NK cells, Neutrophils, Macrophages, DCs	**NK cells:** CD56, proliferation, cytotoxicity↑	**NK cells:** CD2, LFA-1, proliferation, cytotoxicity↑	**NK cells:** CD16, CX3CR, proliferation, cytotoxicity↑	**NK cells:** CD69, NKp46, proliferation, cytotoxicity↑	**NK cells:** Cytotoxicity↑ **DCs:** IFN-α: differentiation in moDCs, HLA-DR, B7, CD11c, CD80, CD83 and CD86↑ IFN-β: ability of mature DCs to stimulate T-cell proliferation and differentiate into Th1↓	**NK cells:** NKp30, proliferation↑ **Macrophages:** Number of $CD14^{high}$ $CD16^{+}$ monocytes; M1 polarization↑	**Neutrophils:** PD-L1, proliferation↑ **Macrophages:** M1 polarization↑ **DCs:** Differentiation in moDCs, MHCII, CD80, CD83, CD86↑
NKT-cells, γδ T-cells	**NKT-cells**: Proliferation↑ **γδ T-cells**: CD69, CD107a, IFN-γ, cytotoxicity↑	NA	**NKT-cells**: Proliferation, cytotoxicity↑	NA	NA	NA	NA
Adoptive immunity cells: Th1, Th2, Th17, Tfh, Tregs, $CD8^{+}$ T-cells, FO B cells, Bregs	**Tregs:** CD25, CTLA4, and HLA-DR, FOXP3, proliferation↑ **$CD8^{+}$ T-cells**: CD25, LAMP-1, proliferation, cytotoxicity↑ PD-1↓	**Th1**: IFN-γ, differentiation from naïve T-cells↑ **Tregs:** The number in TME↓ **$CD8^{+}$ T-cells**: IFN-γ, GrzmB, proliferation, cytotoxicity↑	**Tregs:** No effect **$CD8^{+}$ T-cells**: CD38, HLA-DR, proliferation, cytotoxicity↑	**Th17**: Differentiation from naïve T-cells, IL17, RORγt, IL23R↑ **Tfh:** Proliferation, differentiation, ICOS↑ **Tregs:** Proliferation↓ **FO B cells:** GC B cell proliferation, plasma cell differentiation↑ **Bregs:** B10 cell proliferation, IL10↑	**Th1**: Differentiation↑ **Th2**: Differentiation↓ **Th17**: Differentiation↓ **$CD8^{+}$ T-cells**: CD38, CD25, GrzmB, TRAIL, FasL, IFN-γ, cytotoxicity↑	**Th2**: Proliferation↓ **Tregs:** Proliferation↑	NA

↑ – increase in expression/secretion or some function (activity), ↓ – decrease in expression/secretion or some function (activity).

For instance, Interleukin IL-2 has received approval for metastatic melanoma and RCC (renal cell carcinoma) therapy given that the cytokine boosts T-lymphocyte and natural killer (NK) cell division. A challenge here is its toxicity that has led to scrutiny of tweaking its properties via biotechnology. Among the three different complexes constituting the IL-2 receptor, the medium-affinity receptor (IL-2Rβ) and the high-affinity receptor (also has IL-2Rα) set off the signaling circuits. Research is on to stem the T regulatory (Treg) cell activation given the copious expression of IL-2Rα on these cells to shift the availability of IL-2 to stimulate T lymphocytes and NK cells.

Figure 1 below documents the approach for looking at controlling the toxicity of IL-2 on the basis of quelling the high-affinity IL-2 receptor (L-2Rα, the IL-2Rβ and the common γ chains) while augmenting the medium-affinity receptor (IL-2Rβ and the common γ chains) to alter signal transducer and activator of transcription-3 (STAT3) and STAT5 via Janus kinase-1 (JAK1) and JAK3 (Berraondo et al, 2019: *This article is distributed under the terms of the Creative Commons Attribution 4.0 International License (http://creativecommons.org/licenses/by/4.0/), which permits unrestricted use, distribution, and reproduction in any medium, provided you give appropriate credit to the original author(s) and the source, provide a link to the Creative Commons license, and indicate if changes were made*):

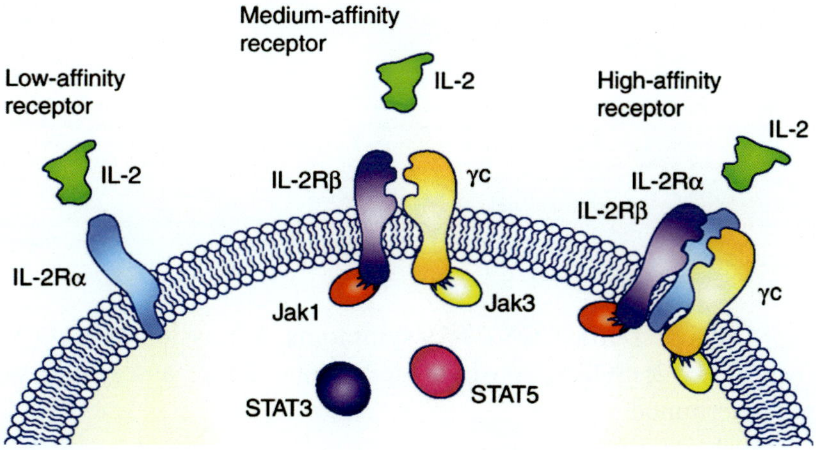

5.1.1 Case Study of Cytokines for COVID-19

A 2021-published report by Dickel and team in *Frontiers in immunology* described as "*the first case report of a patient with severe ARDS from COVID-19 suffering from multiple secondary infections*" probed the impact of Interferon-y (IFNy). Given that earlier work had highlighted how this interferon could boost monocytic human leukocyte antigen receptors (HLA-DR) in order to target secondary infections in septic patients, the researchers administered a one-time INFy injection (100 mcg s.c.) in a 61-year-old man with COVID-19–associated acute respiratory distress syndrome (ARDS), repetitive secondary infection and diminished HLA-DR expression in circulating monocytes (mHLA-DR).

The parameters of the patient are documented below in table 2 below for further understanding following intensive care unit (ICU) admission.

The red in figure 3 below documents immunomodulation while green shows antibiotics for initial *Haemophilus influenza* and later resistant *Pseudomonas aeruginosa* (3-MRGN) and non-resistant *Klebsiella oxytoca*; * is ECMO (venous extracorporeal membrane oxygenation); the 28th day saw the administration of subcutaneous (s.c) 100 mcg IFNy (© 2021 Dickel, Grimm, Amschler, Schnitzler, Schanz, Moerer, Payen, Tampe and Winkler: *This is an open-access article distributed under the terms of the Creative Commons Attribution License (CC BY). The use, distribution or reproduction in other forums is permitted, provided the original author(s) and the copyright owner(s) are credited and that the original publication in this journal is cited, in accordance with accepted academic practice. No use, distribution or reproduction is permitted which does not comply with these terms*):

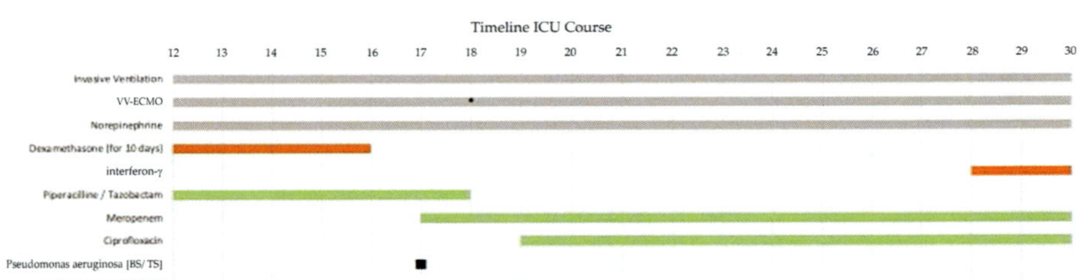

Key: *Days post- first COVID-19 symptoms, Admission to ICU (ARDS center) on day 12; SOFA, sepsis-related organ failure assessment score; N/D, not determined

	After onset of first COVID-19 symptoms*									
	Normal range	**Day 12***	**Day 14**	**Day 16**	**Day 18**	**Day 20**	**Day 22**	**Day 25**	**Day 26**	**Day 29**
Laboratory markers										
Leukocytes ($\times 10^3/\mu L$)	4.0–11.0	15.40	17.50	16.30	11.87	13.10	10.60	8.54	6.99	7.10
CRP (mg/L)	≤5	316.8	351.3	432.4	324.0	N/D	313.9	242.8	239.4	190.7
PCT (μg/L)	<0.07	0.39	0.37	0.47	1.04	0.69	1.15	0.69	0.72	1.43
IL-6 (pg/mL)	<7.0	164.3	146.8	472.6	238.6	262.5	285.9	187.2	72.9	120.4
Ventilation										
Horowitz-Quotient	>300	127.5	97.8	116.9	56.0	65.0	69.0	80.0	57.0	74.6
ECMO										
Flow (1/min)		4.2	4.3	5.0	8.0	7.0	7.7	7.0	7.3	7.3
Clinical Finding										
SOFA-Score		10	11	11	12	16	15	15	16	15
Medication										
Norepinephrine in (μg/kg/min)		0.01	0.04	0.25	0.28	0.25	0.22	0.16	0.11	0.08

Severe acquired immunodepression in terms of diminished HLA-DR (CD14+ mHLA-DR+) within 22.4% to 28.4% was observed on flow-cytometry analysis (FACS). The "off-label use of subcutaneously (s.c.) INFy in order to attenuate the acquired immunosuppression" resulted in an increase of HLA-DR+/CD 14+ cells by 56.4% as depicted in below (however refractory ARDS caused the mortality of the patient on day 29) warrants detailed further investigation.

Figure 4 documents lung x-rays on day 12 (A) and on day 28 (B); FACS for monocytic human leukocyte antigen receptors (mHLA-DR) on monocytes (CD 14+) prior to IFNy treatment on day 26 (C) and post-IFNy treatment on day 29 (D) (© 2021 Dickel, Grimm, Amschler, Schnitzler, Schanz, Moerer, Payen, Tampe and Winkler: *This is an open-access article distributed under the terms of the Creative Commons Attribution License (CC BY). The use, distribution or reproduction in other forums is permitted, provided the original author(s) and the copyright owner(s) are credited and that the original publication in this journal is cited, in accordance with accepted academic practice. No use, distribution or reproduction is permitted which does not comply with these terms*):

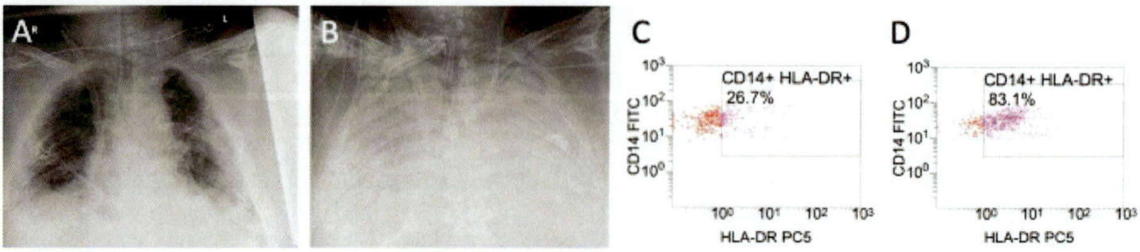

Parameters	Day 16*	Day 18	Day 26	Day 29
Monocytes (cells/μL)	378	203	125	111
CD14 + (cells/μL)	325 (85.3%)	160 (78.9%)	97 (78.0%)	91 (82.0%)
CD14 + HLA-DR + (cells/μL)	73 (22.4%)	45 (28.4)	26 (26.3%)	75 (82.2%)
Lymphocytes (cells/μL)	1699	1210	1507	1181
CD3 + (cells/μL)	875 (51.5%)	734 (60.7)	1014 (67.3%)	858 (72.6%)
HLA-DR + (cells/μL)	126 (14.4%)	67 (9.1%)	87 (8.6%)	76 (8.9%)

5.1.2 Case Study of Cytokines in Cat Diseases

Research has probed the impact of low-dose oral human interferon-α (HuIFN-α) on cats with feline immunodeficiency virus (FIV) infection manifesting symptoms on the lines of AIDS. While the cats treated with interferon documented a copious increase in survival against the placebo, the immune response in the HuIFN-α-administered animals was "better" in terms of opportunistic infection control with augmented CD8+ T cell numbers. Further, another study documented augmented survival in cats infected with feline leukemia virus (FeLV) administered low-dose HuIFN-α as opposed to the controls. Detailed elucidation and scrutiny of the utility of cytokine administration can facilitate the quantitation of appropriate dosage and usage of this approach in pets and farm animals (Frazzini, Riva and Amadori, 2021)

Figure 5 below illustrates the impact of cytokine therapy in cats (a) opportunistic in a FIV-infected animal prior to IFN-α treatment (b) Lowered lesions post IFN-α treatment (Results of *Pedretti E, Passeri B, Amadori M, Isola P, Di Pede P, Telera A, Vescovini R, Quintavalla F, Pistello M. Low-dose interferon-alpha treatment for feline immunodeficiency virus infection.*

Vet Immunol Immunopathol. 2006 Feb 15; 109(3-4):245-54: Frazzini, S., Riva, F., & Amadori, M. (2021): © 2021 by the authors. Licensee MDPI, Basel, Switzerland. *This article is an open access article distributed under the terms and conditions of the Creative Commons Attribution (CC BY) license: https://creativecommons.org/licenses/by/4.0/*):

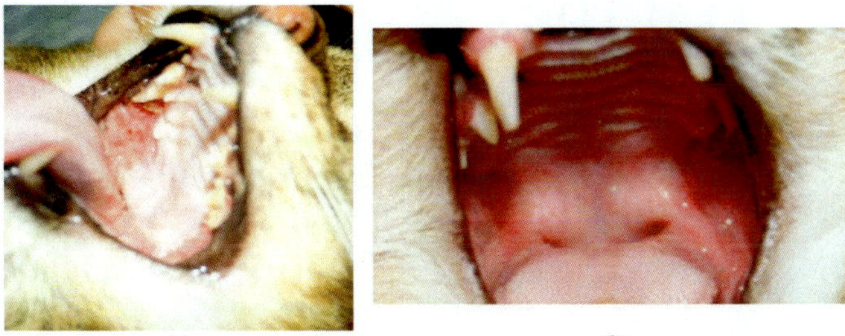

(a) (b)

The pleiotropic effect of cytokines i.e. the regulation of many immune cells by each cytokine emerges as the hallmark of cytokines. The toxicity and short half-lives have initiated a quest to diminish these responses while also combining cytokines with other approaches. Further, specific targeting of these molecules to only the "site" can circumvent the systemic toxicity facilitated by mAbs or gene therapies (Berraondo et al, 2019; Chulpanova et al, 2020). The future of cytokine therapy is depicted in figure 6 below. Combination with TILs tumor-infiltrating lymphocytes, CARs chimeric antigen receptor T cells and ADCC antibody-dependent cellular cytotoxicity is depicted for immunostimulatory cytokines (left) and immunosuppressive cytokines (right) (Berraondo et al, 2019: *This article is distributed under the terms of the Creative Commons Attribution 4.0 International License (http://creativecommons.org/licenses/by/4.0/), which permits unrestricted use, distribution, and reproduction in any medium, provided you give appropriate credit to the original author(s) and the source, provide a link to the Creative Commons license, and indicate if changes were made*):

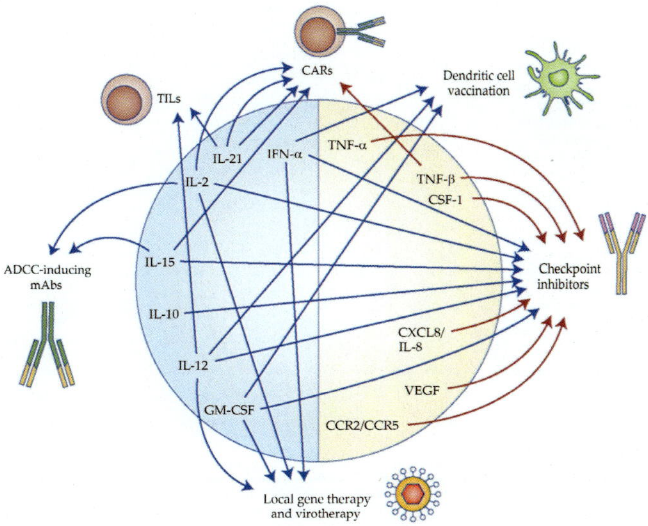

5.2 GROWTH FACTORS

The molecules vitally involved in growth, proliferation and morphogenesis called growth factors (GFs) have seen studies resulting in their use in the clinic for several conditions. Table 3 below outlines growth factors employed in regenerative medicine (© 2020 Ren, Zhao, Lash, Martino and Julier. *This is an open-access article distributed under the terms of the Creative Commons Attribution License (CC BY). The use, distribution or reproduction in other forums is permitted, provided the original author(s) and the copyright owner(s) are credited and that the original publication in this journal is cited, in accordance with accepted academic practice. No use, distribution or reproduction is permitted which does not comply with these terms*):

Product	GF	Delivery system	Target tissue/disease	Approved authority	References
Augment® Bone Graft	PDGF-BB	Beta-tricalcium phosphate	Ankle fusion, hindfoot	FDA	FDA, 2015a
Increlex®	IGF-1	Subcutaneous Injection	Primary IGF-1 deficiency	FDA	FDA, 2005; National Drug Strategy, 2006
Infuse® Bone Graft	BMP-2	Collagen sponge	Spinal fusion, bone regeneration	FDA	James et al., 2016
Kepivance®	FGF-7 (KGF)	i.v. injection	Gastrointestinal injury	FDA	FDA, 2015b
OP-1® Putty	BMP-7	Bovine bone-derived collagen	Spinal fusion, bone regeneration	FDA	Okabe et al., 2013
PELNAC®	FGF-2 (bFGF)	Collagen sponge	Bedsores, cutaneous ulcers	Pharmaceuticals and Medical Devices Agency (Japan)	Kakudo et al., 2019
REGEN-D®	EGF	Cellulose gel	Foot ulcer	Ministry of Food and Drug Safety (South Korea)	Frew et al., 2007
Regranex®	PDGF-BB	Sodium carboxymethylcellulose-based topical gel	Chronic diabetic wound	FDA	FDA, 2008
Citrix® CRS	TGF-β1	Topical	Aged skin		Aldag et al., 2016

BMP, bone morphogenetic protein; EGF, epidermal growth factor; FGF, fibroblast growth factor; HGF, hepatocyte growth factor; IGF, insulin-like growth factor; PDGF, platelet-derived growth factor; TGF, transforming growth factor; FDA, U.S. Food Drug Administration.

GFs	Desired function(s)	Half-life in blood	Side-effect(s) in humans	References
BMP-2	Osteogenic factor	1–4 h	Ectopic bone formation, abnormal osteogenesis, inflammatory complications, urogenital events, wound complications, increase cancer risk	Carragee et al., 2013; Carreira et al., 2014; James et al., 2016
BMP-7	Osteogenic factor, regulate proliferation of neural progenitor cells	1–4 h	Not reported	Calori et al., 2009; Carreira et al., 2014; Kowtharapu et al., 2018
EGF	Stimulates proliferation and differentiation of epithelial cells	<1 min	Not reported	Mitchell et al., 2016
FGF-2	Stimulates proliferation and differentiation of various cell types, angiogenesis	7.6 h	Not reported	Beenken and Mohammadi, 2009; Maddaluno et al., 2017
HGF	Stimulates epithelial cell proliferation and morphogenesis, angiogenesis	3–5 min	Not reported	Yu et al., 2007; Nakamura et al., 2019
IGF-1	Enhances neuronal growth, myelination, endometrial epithelial cell proliferation, inhibition of cell apoptosis	3–5 h	Not reported	Leroith et al., 1992; Wang et al., 2018
PDGF-BB	Proliferation of various cell types, extracellular matrix synthesis, vascularization	30 min	Increase cancer risk	Jin et al., 2008; Saika et al., 2011; Mao and Mooney, 2015; Yamakawa and Hayashida, 2019
FGF-7 (KGF)	Epithelium morphogenesis, re-epithelialization	4–6 h	Enhance epithelial tumor cell growth	FDA, 2004a,b; Yamakawa and Hayashida, 2019
TGF-β1	Differentiation of bone-forming cells, antiproliferative factor for epithelial cells	>100 min (latent form) 2–3 min (active form)	Not reported	Hermonat et al., 2007; Lee et al., 2011; Tian et al., 2011
VEGF-A	Angiogenic factor	30 min	Edema, systemic hypotension	Simons and Ware, 2003; Stefanini et al., 2008; Yamakawa and Hayashida, 2019

BMP, bone morphogenetic protein; EGF, epidermal growth factor; FGF, fibroblast growth factor; HGF, hepatocyte growth factor; IGF, insulin-like growth factor; PDGF, platelet-derived growth factor; TGF, transforming growth factor; VEGF, vascular endothelial growth factor.

As an example, Koike and team (2021) reported that wound closure in mice was accelerated by fibroblast growth factor (FGF) 2 via epithelial–mesenchymal transition (EMT) of keratinocytes. Wounds on C57BL/6 mice closed in 7.8 ± 0.2 days when exposed to FGF2 as opposed to 9.8 ± 0.3 days in vehicle-treated controls. The wound edge also documented migrating keratinocytes with intermediate filaments, vimentin and other molecules associated with EMT associated with new tissue formation and wound healing.

Figure 7 below illustrates the effect of FGF-2 on murine wounds: (**a**) smaller wounds on FGF2-treatment (**b**) Lesser days for wound closure by FGF2 treatment. (**c**,**d**) A more thickened epidermis (double arrow) in FGF2-treated wounds vs. vehicle-treated wounds. Scale bar: 40 µm. (**e**) The wound edge documenting spindle-shaped keratinocytes (arrow heads) in FGF2-treated wounds (higher magnification images at the corner). Scale bar: 20 µm. (**f**) Keratinocytes were the migrating cells with anti-pan keratin antibody staining. Scale bar: 40 µm. (**g**) More migrating keratinocytes in FGF2-treated wounds vs. control wounds: FGF2: fibroblast growth factor 2, HE: hematoxylin and eosin staining (Koike et al, 2021: *Open Access: This article is licensed under a Creative Commons Attribution 4.0 International License, which permits use, sharing, adaptation, distribution and reproduction in any medium or format, as long as you give appropriate credit to the original author(s) and the source, provide a link to the Creative Commons licence, and indicate if changes were made. The images or other third-party material in this article are included in the article's Creative Commons licence, unless indicated otherwise in a credit line to the material. If material is not included in the article's Creative Commons licence and your intended use is not permitted by statutory regulation or exceeds the permitted use, you will need to obtain permission directly from the copyright holder. To view a copy of this licence, visit http://creativecommons.org/licenses/by/4.0/*):

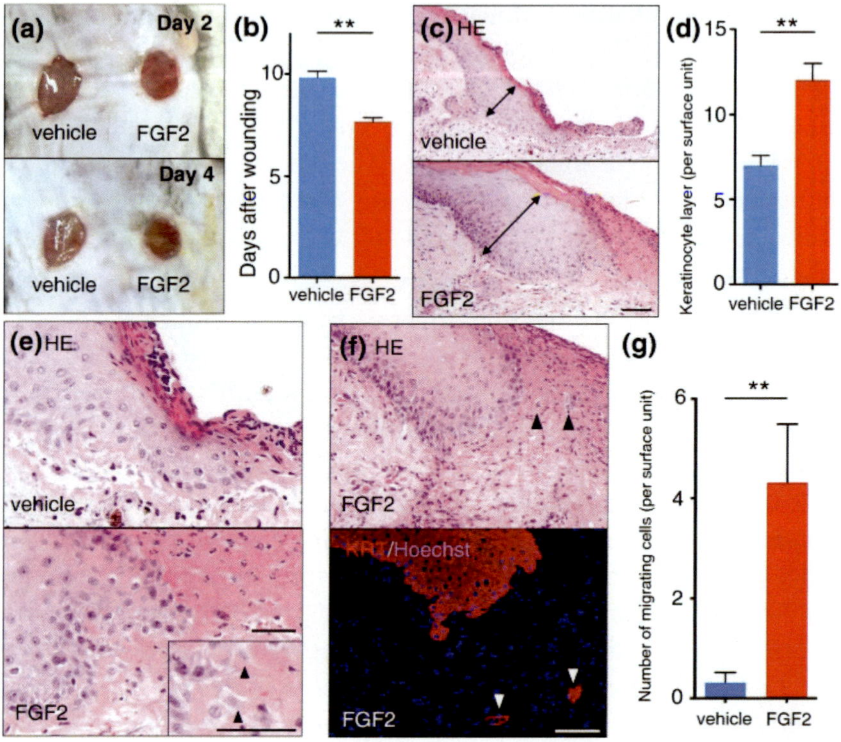

A meta-analysis of six studies involving 530 patients by Bui and team (2019) in the *International Journal of environmental research and public health* documented the promise of recombinant human epidermal growth factor (rhEGF) in the healing of diabetic foot ulcers (DFUs). While the evident healing of such ulcers was facilitated by the growth factor, it is vital that the management of diabetes warrants a multi-pronged approach.

Table 4 below documents the effect of EGF on DFUs (Bui et al, 2019; *International journal of environmental research and public health* © 2019 by the authors. Licensee MDPI, Basel, Switzerland. *This article is an open access article distributed under the terms and conditions of the Creative Commons Attribution (CC BY) license: http://creativecommons.org/licenses/by/4.0/*):

Key: rhEGF (recombinant human epidermal growth factor); ABI (ankle-brachial index); TcPO2 (transcutaneous oxygen pressure); SD (standard deviation); NA (not available).

Author	Intervention Type	Route	Apply Frequency	Treatment Duration (weeks)	Ulcer Data						
					Location	Severity	Ischemia Level	Infection	Ulcer Area ± SD (cm²)	Ulcer Duration ± SD (Weeks)	Complete Healing Rate (%)
Tsang 2003	Actovecgin %	Topical	Once a day	12	Foot	Wagner 1 or 2	ABI ≥ 0.7	NA	3.48 ± 0.82	12 ± 15.47	42.1
	0.02 wt% rhEGF				Foot				3.4 ± 1.1	11.48 ± 14.68	57.14
	0.04 wt% rhEGF				Foot				2.78 ± 0.82	8.24 ± 5.55	95.3
Afshari 2005	1000 mg 1%sulfadiazine	Topical	Once a day	8	Foot	Wagner 1 or 2	ABI < 1: 49% participants	NA	103.4 ± 147.8	59.7 ± 55.5	10
	1 mg rhEGF in 1000 mg 1% sulfadiazine				Foot		ABI < 1: 50% participants		87.5 ± 103.2	42.9 ± 38.4	23.3
Fernández-Montequín 2009	placebo	Intralesional injection	Three times a week on alternate days	8	Foot	Wagner 3 or 4	Hemoglobin ≥100 g/L	NA	21.8	4.9	52.1
	25 µ/g				Foot				4.3	20.1	52.1
	75 µ/g				Foot				4.3	28.5	75.5
Gomez-Villa 2014	Placebo	Intralesional injection	Three times a week on alternate days	8	Foot	Texas 1, 2, and 3	ABI > 0.6	NA	11.9 ± 11.8	15.3 ± 8.4	0
	75 µ/g				Foot				19.2 ± 15.7	25.8 ± 44	23.5
Singla 2014	Betadine	Topical	Once every two weeks	8	Foot	Wagner 1 or 2	ABI ≥ 0.75	NA	NA	NA	12
	Urogastrone (rhEGF) gel 15 g				Foot				NA	NA	48
Park 2018	Saline	Topical	Twice a day	12	Foot	Wagner 1 or 2	TcPO2 ≥ 30 mmHg or palpable dorsalis pedis artery or posterior tibial artery	NA	2.35 ± 2.69	29.6 ± 60.2	50.6
	0.005% rhEGF spray				Foot				2.8 ± 3.72	38.48 ± 70.24	73.2
Xu 2018	Saline	Topical	Once a day	8	Foot	Wagner 1 or 2	NA	NA	2.35 ± 2.69	29.6 ± 60.2	NA
	40 IU/cm²				Foot				2.8 ± 3.72	70.24 ± 38.48	NA

While the vital involvement of growth factors in healing makes them get the limelight, copious clinical translations are yet to see the light of the day. This warrants novel delivery systems to be explored. Such systems can entail employing suitable biomaterials to immobilize growth factors via affinity binding, covalent or enzyme conjugation or augmenting their thermal and protease resistance and altering the sequences for synergistic functioning.

Figure 8 below illustrates the engineering of growth factors via (A) biomaterials (B) stability and signaling (C) for augmented functioning (© 2020 Ren, Zhao, Lash, Martino and Julier. *This is an open-access article distributed under the terms of the Creative Commons Attribution License (CC BY). The use, distribution or reproduction in other forums is permitted, provided the original author(s) and the copyright owner(s) are credited and that the original publication in this journal is cited, in accordance with accepted academic practice. No use, distribution or reproduction is permitted which does not comply with these terms*):

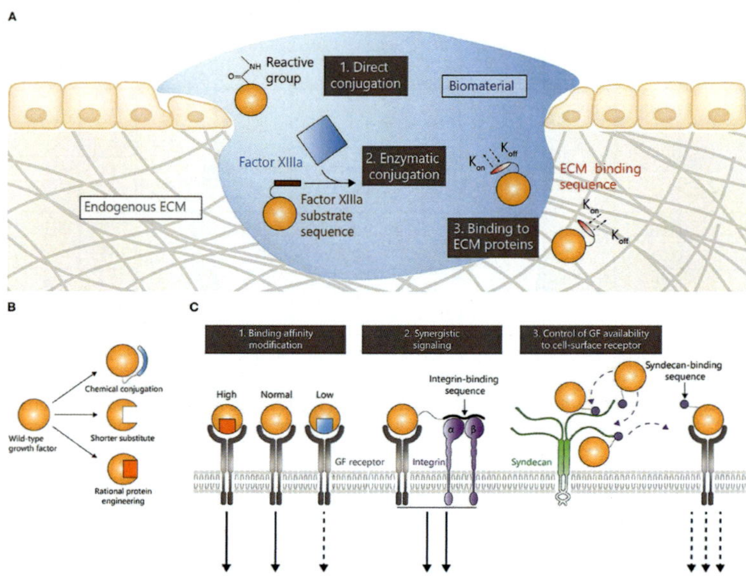

5.3 NANOBODIES

First obtained in 1993 from the serum of the camel (*Camelus dromedarius*), nanobodies lack the light chain and are heavy-chain dimmers. Their dimensions fetched them their name by the Belgian company Ablynx®. The complementarity-determining region 3 (CDR3) of nanobodies is 3 to 28 amino acids over 8 to 15 amino acids of the VH domain (variable domain of the heavy immunoglobulin chain) hence, not compromising their ability to bind antigens. Their ability to form finger-like structures facilitates hidden epitopes not accessible to mAbs while paving the way for the quest of new active sites or receptor pockets. Amino acid engineering can augment the ability of nanobodies to remain stable and soluble in water (Bao et al, 2021).

Figure 9 below illustrates ***TOP***: (a) nanobodies for imaging and therapy (b) A typical mAb has 2 light (L) chains and heavy (H) chains with 2 functional domains, i.e., variable region (VR) and one constant region (CR) (c) HcAb naturally is without light chains and CH1 domains. The nanobody is its variable fragment (d) Nanobody with 4 framework regions and 3 CDRs (e) Producing bivalent nanobodies with a third nanobody to bind serum albumin (anti-Alb) to augment the half-life (Bao et al, 2021: *Open Access: This article is licensed under a Creative Commons Attribution 4.0 International License, which permits use, sharing, adaptation, distribution and reproduction in any medium or format, as long as you give appropriate credit to the original author(s) and the source, provide a link to the Creative Commons licence, and indicate if changes were made. The images or other third party material in this article are included in the article's Creative Commons licence, unless indicated otherwise in a credit line to the material. If material is not included in the article's Creative Commons licence and your intended use is not permitted by statutory regulation or exceeds the permitted use, you will need to obtain permission directly from the copyright holder. To view a copy of this licence, visit http://creativecommons.org/licenses/by/4.0/*):

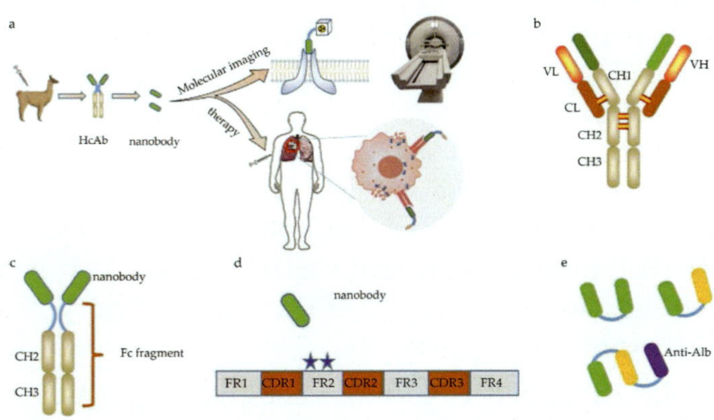

BOTTOM: Nanobodies for specifically targeting diseases. This can entail (a) radiopharmaceuticals with a radionuclide (b) as conjugated drug carriers for faster clearance and lower "off-targets" (c) conjugation with immune toxins (d) forming fusion peptides to circumvent the lack of the Fc region (Bao et al, 2021: *Open Access: This article is licensed under a Creative Commons Attribution 4.0 International License, which permits use, sharing, adaptation, distribution and reproduction in any medium or format, as long as you give appropriate credit to the original author(s) and*

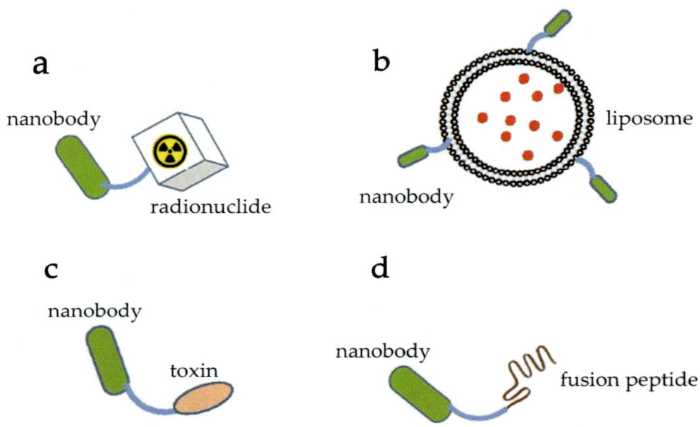

Among the various techniques explored to battle severe acute respiratory syndrome coronavirus 2 (SARS-CoV-2), nanobodies too have shown their mettle. For example, Koenig and team (2021) reported the development of nanobodies to target the virus in *Science*. The work entailed the production of variable domains of camelid heavy-chain–only antibodies (VHHs) from an alpaca and a llama with the receptor binding domains (RBD) of the SARS-CoV-2 spike protein. Four nanobodies emerged as potent from other candidates: VHH E from the llama and VHHs U, V, and W from the alpaca with VHH E as the most promising [half-maximal inhibitory concentration (IC_{50}) value of 60 nM].

Figure 10 below depicts Cryo-EM reconstructions [(A) and (C)] and atomic models [(B) and (D)] of VHH E [(A) and (B)] and VHH V [(C) and (D)] with the trimeric SARS-CoV-2 spike. Binding of VHH E (in green) to SARS-CoV-2 in a 3-up conformation; the resolution is 3.3 Å [0.143 Fourier shell correlation (FSC)]. (C and D) Binding of VHH V (in red) to SARS-CoV-2 in a 2-up conformation with all VHH binding sites occupied at a resolution of 3.0 Å (0.143 FSC). In the most abundant complex, VHH V binds to the RBD in the up or the down conformation. (E and F) Microscopy of HEK 293 cells expressing SARS-CoV-2 S Δ18 treated with 1 μM of the indicated VHHs (E) Quantifying fusion (F) Representative images of cells 12 hours post-treatment: Scale bars, 100 μm (Koenig et al, *Science*, 2021:

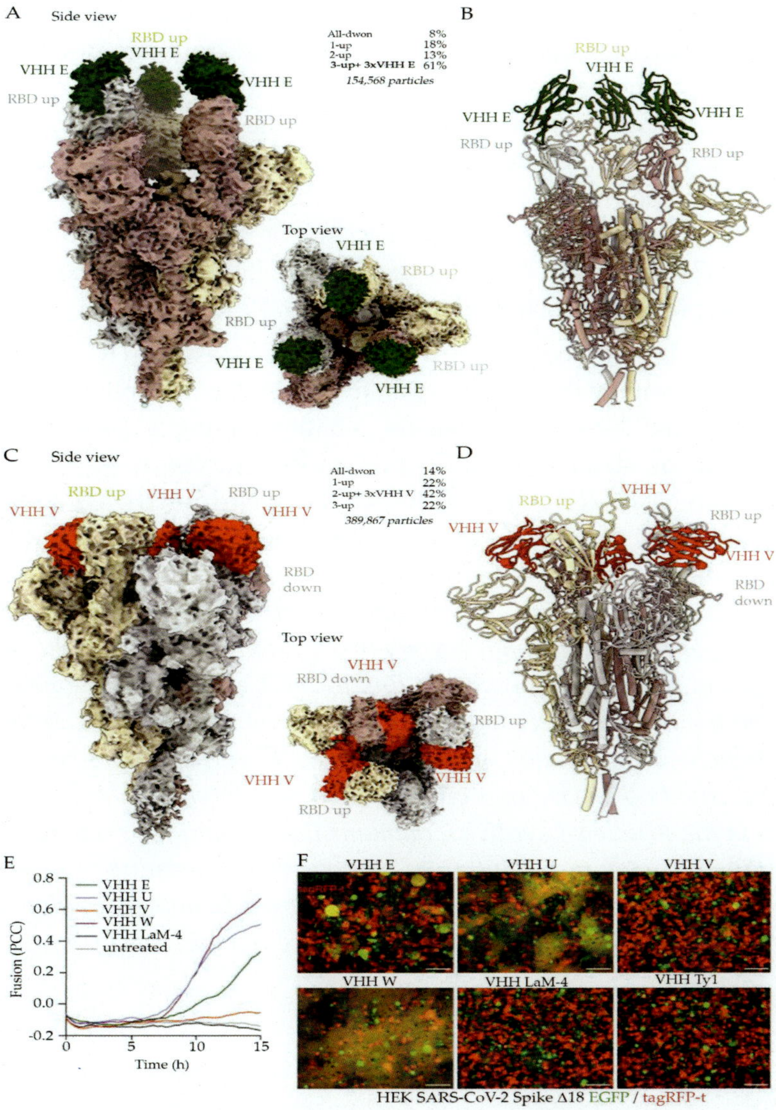

Further analyses revealed SARS-CoV-2 mNeonGreen replication inhibition in human cells by the individual as well as mixtures of nanobodies: additive inhibition emerged where two nanobody mixes at half-concentration was as effective as only the individual nanobodies at the full concentration in neutralizing the virus. Given the emergence of mutants, this was quelled by the simultaneous targeting of two neutralizing epitopes with no mutational escape recorded for biparatopic VHHs EV and VE paving the way for further elucidation (Koenig et al, 2021).

In another 2021-published article by Xu et al in *Nature* documented the promise of anti-RBD nanobodies from llamas and from mice. Two groups of nanobodies emerged with one targeting a rarely targeted and highly conserved RBD region of coronaviruses with the other group targeting the RBD–ACE2 interface. When the nanobodies of the second group were expressed as homotrimers, they could neutralize SARS-CoV-2 variants with E484K substitution (in the B.1.351 variant) or N501Y substitution (seen in the B.1.1.7 variant and the P.1 variant with N501Y, K417T and E484K) showing the promise of nanobodies in COVID-19 treatment.

Figure 11 below documents **a** Neutralization assays (IC_{50} values) of pseudoviruses with wild-type or mutant SARS-CoV-2 spike **b** Trivalent Nb15, Nb56 and Nb12 and bivalent Nb30 targeting **c** Biosensor-based measurement of nanobody–RBD immunocomplexes **d**–**g**, Binding of nanobodies to Nb12–RBD (**d**), Nb30–RBD (**e**), Nb15–RBD (**f**) and Nb56–RBD (**g**) immunocomplexes (Xu et al, 2021: *Open Access: This article is licensed under a Creative Commons Attribution 4.0 International License, which permits use, sharing, adaptation, distribution and reproduction in any medium or format, as long as you give appropriate credit to the original author(s) and the source, provide a link to the Creative Commons license, and indicate if changes were made. The images or other third party material in this article are included in the article's Creative Commons license, unless indicated otherwise in a credit line to the material. If material is not included in the article's Creative Commons license and your intended use is not permitted by statutory regulation or exceeds the permitted use, you will need to obtain permission directly from the copyright holder. To view a copy of this license, visit http://creativecommons.org/licenses/by/4.0/):*

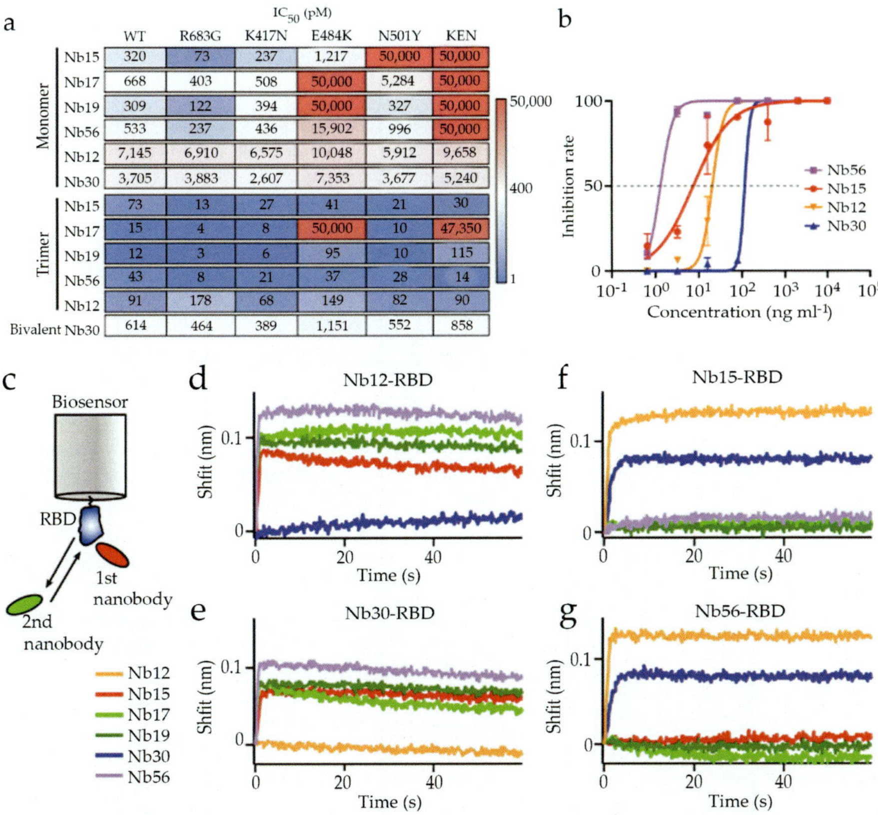

Table 5 below documents some of the research work done on nanobodies in terms of therapy (Bao et al, 2021: *Open Access: This article is licensed under a Creative Commons Attribution 4.0 International License, which permits use, sharing, adaptation, distribution and reproduction in any medium or format, as long as you give appropriate credit to the original author(s) and the source, provide a link to the Creative Commons licence, and indicate if changes were made. The images or other third party material in this article are included in the article's Creative Commons licence, unless indicated otherwise in a credit line to the material. If material is not included in the article's Creative Commons licence and your intended use is not permitted by statutory regulation or exceeds the permitted use, you will need to obtain permission directly from the copyright holder. To view a copy of this licence, visit http://creativecommons.org/licenses/by/4.0/*):

Function and target	Combined with	Advantages	Disadvantages	Solution(s)
Antagonist: EGFR, CXCR4, P2X7, HGF	-	(i) Small size (ii) High solubility and stability (iii) Excellent tissue penetration (iv) Recognizing new targets (v) Using together with mAbs	(i) Fast blood clearance (ii) High renal uptake (iii) Lack of Fc (iv) Immunogenicity	(i) Constructing multivalent nanobody (ii) Co-injecting with cationic amino acids (iii) Conjugating with an anti-albumin unit (iv) Linking with effector domains
Nanobody-based radionuclide: HER2	^{225}Ac, ^{131}I, ^{177}Lu	(i) Fast blood clearance (ii) Suited for conjugation	(i) High renal uptake (ii) Radiation toxicity to healthy cells	(i) Constructing multivalent nanobody, co-injecting with cationic amino acids, conjugating with an anti-albumin unit (ii) Linking with residualizing prosthetic group such as SGMIB (iii) Selecting high-affinity and high-internalization nanobody
Nanobody-mediated drug delivery system: EGFR, HER2	(i) Pharmaceutic carriers (ii) Chemotherapeutic drugs	(i) Can act as antagonist itself (ii) High specificity (iii) Suited for conjugation	(i) Fast blood clearance (ii) The drug can damage normal cells	(i) Encapsulation in carriers, (ii) PEGylation
Nanobody-based immunotoxin: EGFR, CD7	(i) Plant toxins (ii) Bacterial protein toxins such as PE and DT	(i) Lethal to cells in all phases (ii) High efficacy	(i) Immunogenicity (ii) Lysosome-sensitive sites in toxin part	(i) Linking nanobody with humanized nanobody scaffold (ii) Deleting lysosome-sensitive sites
Nanobody-peptide fusions: EGFR, DR, CEA	(i) The ligand of death receptor (TRAIL) (ii) Fc domains (iii) Cytokine	(i) Inducing ADCC and CDC (ii) Specifically recruiting effector cells to lesions	Fast blood clearance	(i) Constructing multivalent or nanobody, (ii) Glycosylation modification (iii) Crucial amino acid mutation in FR2

5.4 RNA AND DNA

While traditional drug design is aimed at targeting proteins, recombinant DNA tools have paved the way for nucleic-acid-based strategies that circumvent the challenge of the former such as the requisite for correct folding and post-translational modifications. *DNA drugs* entail the use of plasmid or viral vectors for nuclear delivery and subsequent expression. Research ubiquitously reports the use of adeno-associated virus (AAV) for this purpose with the risk of integration considered as minimal. An example is Luxturna that encodes the RPE65 gene in AAV (serotype 2) for the restoration of diminished levels of the protein seen in inherited retinal dystrophy (FDA approval in 2018 and EU approval in 2018) (Bajan and Hutvagner, 2020; Damase, Sukhovershin, Boada, Taraballi, Pettigrew and Cooke, 2021).

Looking at the rate to target COVID-19, figure 12 below summarizes the various approaches being scrutinized inclusive of DNA vaccines (Kyriakidis et al, 2021: *Open Access: This article is licensed under a Creative Commons Attribution 4.0 International License, which permits use, sharing, adaptation, distribution and reproduction in any medium or format, as long as you give appropriate credit to the original author(s) and the source, provide a link to the Creative Commons license, and indicate if changes were made. The images or other third party material in this article are included in the article's Creative Commons license, unless indicated otherwise in a credit line to the material. If material is not included in the article's Creative Commons license and your intended use is not permitted by statutory regulation or exceeds the permitted use, you will need to obtain permission directly from the copyright holder. To view a copy of this license, visit http://creativecommons.org/licenses/by/4.0/*):

A Attenuated live pathogen vaccines entail lowering the pathogen virulence for eliciting memory responses. B Inactivated pathogen vaccines comprise whole pathogens exposed to heat or chemical treatment. C Subunit vaccines contain purified antigens from cell cultures or recombinant expression. D Virus-like particles entail recombinant expression and assembly. E Viral vector vaccines entail the use of a viral vector for a foreign antigen. F and G Nucleic acid (DNA and mRNA) vaccines entail the administration of the nucleic acid(s) of an immunogenic protein of the pathogen.

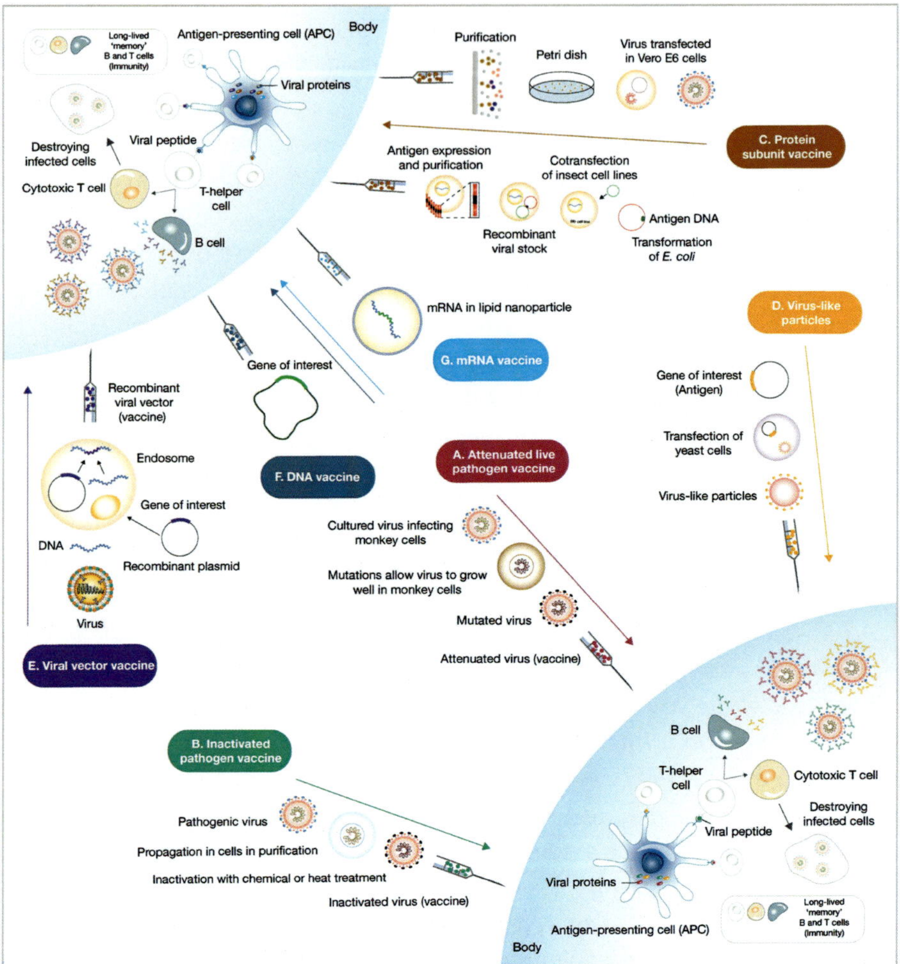

There are *CRIPR-Cas-based drugs* under evaluation inclusive of CTX001 (CRISPR Therapeutics) targeting the BCL11A enhancer to boost γ-globin as a substitute of defective adult hemoglobin (HbA) in sickle cell disease and beta-thalassemia and EDIT-101 (Editas Medicine) aimed at repairing the phototransduction gene *CEP290* in Leber Congenital Amaurosis type 10 (LCA10), to name a few (Damase, Sukhovershin, Boada, Taraballi, Pettigrew and Cooke, 2021).

A slew of *RNA therapeutics* is also being pursued in research: the various categories and molecules available or under scrutiny are presented in figure 13 and table 6 below, respectively:

RNA therapeutics: types: (© 2021 Damase, Sukhovershin, Boada, Taraballi, Pettigrew and Cooke. *This is an open-access article distributed* *under the terms of the Creative Commons Attribution License (CC BY). The use, distribution or reproduction in other forums is permitted, provided the original author(s) and the copyright owner(s) are credited and that the original publication in this journal is cited, in accordance with accepted academic practice. No use, distribution or reproduction is permitted which does not comply with these terms*):

Key: ASO, antisense oligonucleotide; RNA, ribonucleic acid; RNAi, RNA interference; siRNA, small interfering RNA; miRNA, microRNA; mRNA, messenger RNA; A, adenosine molecule; AAAAA, poly A tail.

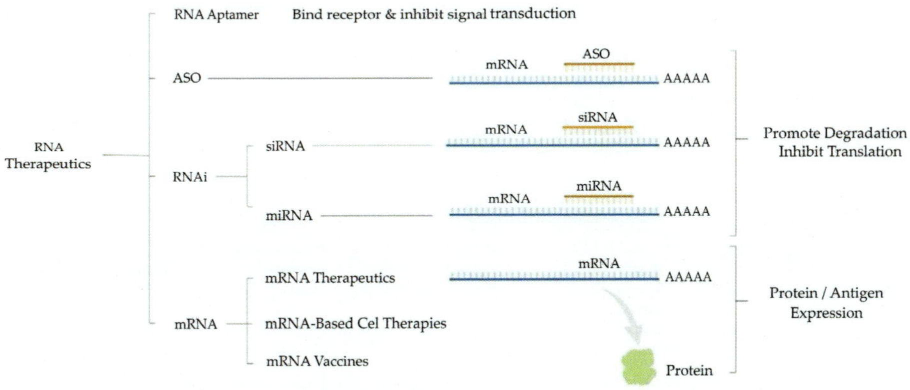

RNA therapeutics available/under probing: (© 2021 Damase, Sukhovershin, Boada, Taraballi, Pettigrew and Cooke. *This is an open-access article distributed under the terms of the Creative Commons Attribution License (CC BY). The use, distribution or reproduction in other forums is permitted, provided the original author(s) and the copyright owner(s) are credited and that the original publication in this journal is cited, in accordance with accepted academic practice. No use, distribution or reproduction is permitted which does not comply with these terms*):

Drug	Type of RNA	Company	Route of delivery	Condition/Disease	Status
Nusinersen (Spinraza)	ASO	Ionis	Intrathecal	Spinal muscular atrophy	FDA approval in 2016
Eteplirsen (Exondys 51)	ASO	Sarepta	Intravenous	Duchenne muscular dystrophy	FDA approval in 2016
Inotersen (Tegsedi)	ASO	Ionis	Subcutaneous	Familial amyloid polyneuropathy	FDA approval in 2018
Volanesorsen (Waylivra)	ASO	Ionis	Subcutaneous	Familial chylomicronemia syndrome	EU approval in 2019
Patisiran (Onpattro)	siRNA	Alnylam	Intravenous	Polyneuropathy	FDA approval in 2018
Givosiran (Givlaari)	siRNA	Alnylam	Subcutaneous	Acute hepatic porphyria	FDA approval in 2019
Cobomarsen (MRG-106)	miRNA	miRagen (Viridian)	Intravenous/subcutaneous	Blood cancers	Phase II
Remlarsen (MRG-201)	miRNA	miRagen (Viridian)	Intradermal	Keloids	Phase II
MRG-110	miRNA	miRagen (Viridian)	Intradermal	Tissue Repair	Phase I
Pegaptanib (Macugen)	Aptamer(RNA)	Bausch + Lomb	Intravitreal	Macular Degeneration	FDA approval in 2014
Emapticap pegol (NOX-E36)	Aptamer(RNA)	NOXXON	Intravenous/Subcutaneous	Diabetic nephropathy, lung and pancreatic cancer	Phase II
Olaptesed pegol (NOX-A12)	Aptamer(RNA)	NOXXON	Intravenous	Brain cancer	Phase I/II
BNT162b2	mRNA	BioNTech and Pfizer	Intramuscular	COVID-19	FDA authorization for emergency use in 2020
mRNA-1273	mRNA	Moderna	Intramuscular	COVID-19	FDA authorization for emergency use in 2020
CVnCoV	mRNA	CureVac	Intramuscular	COVID-19	Phase III
AZD8601	mRNA	Moderna/Astrazeneca	Epicardial	Ischemic heart disease	Phase II
mRNA-1647	mRNA	Moderna	Intramuscular	Cytomegalovirus infection	Phase II
P-BCMA-101	mRNA	Poseida	Intravenous	Multiple myeloma	Phase II
mRNA-4157	mRNA	Moderna	Intramuscular	Cancer	Phase II
mRNA-3704	mRNA	Moderna	Intravenous	Methylmalonic aciduria	Phase I/II
MRT5005	mRNA	Translate Bio	Inhalation	Cystic Fibrosis	Phase I/II
mRNA-2416	mRNA	Moderna	Intratumoral	Solid tumors/ lymphoma/advanced ovarian carcinoma	Phase I/II
BNT131 (SAR441000)	mRNA	BioNTech/ Sanofi/Genmab	Intratumoral	Advanced melanoma	Phase I/II
Descartes-08	mRNA	Cartesian	Intravenous	Generalized myasthenia gravis	Phase I/II
BNT122	mRNA	BioNTech	Intravenous	Solid tumors	Phase I/II
mRNA-2752	mRNA	Moderna	Intratumoral	Solid tumors	Phase I
MEDI1191	mRNA	Moderna	Intratumoral	Solid tumors	Phase I
mRNA-1944	mRNA	Moderna	Intravenous	Chikungunya infection	Phase I
CV8102	mRNA	CureVac	Intratumoral	Solid tumors	Phase I
ARCT-810	mRNA	Arcturus	Intravenously	Urea disorder	Phase I
CV7202	mRNA	CureVac	Intramuscular	Rabies	Phase I
mRNA-1893	mRNA	Moderna	Intramuscular	Zika	Phase I
CV9202	mRNA	CureVac	Intradermal	Non-small cell lung cancer	Phase I
mRNA-5671	mRNA	Moderna	Intravenous	Cancer	Phase I
BNT111	mRNA	BioNTech	Intravenous	Advanced Melanoma	Phase I

The table summarizes information available at ClinicalTrials.gov (https://clinicaltrials.gov/) as of January 26, 2021. ASO, antisense oligonucleotide; siRNA, small interfering RNA; miRNA, microRNA; RNA, ribonucleic acid; mRNA, messenger RNA.

Let us discuss two approved therapeutics, the first is aimed at spinal muscular atrophy (SMA) arising from diminished survival motor neuron (SMN) protein levels on account of ubiquitously *SMN2* without exon 7. The drug Nusinersen (Spinraza™, Biogen) has received approval for SMA therapy and is an ASO binding to intron 7 of the SMN2 pre-mRNA molecule to displace the masking heterogeneous nuclear riboprotein (hnRNP A1/2) and allow exon 7 to be included to augment functional SMN levels. Another drug is the first short interfering RNA (siRNA) drug called Onpattro (patisiran) (Alnylam Pharmaceuticals) targeting hereditary transthyretin-mediated (hATTR) amyloidosis. Transthyretin gene (*TTR)* mutations cause misfolded TTR protein that warrants liver transplantation. The binding of the siRNA to the 3′ UTR of the TTR transcript causes its degradation to lower TTR protein synthesis to diminish amyloidosis: the safety of the drug and augmented quality of life and clinical outcome in patients has led to its approval (Bajan and Hutvagner, 2020).

Figure 14 below illustrates two approved RNA therapeutics: (A) Nusinersen removes the masking by hnRNP on SMN2 resulting in the SMN2Δ7 by displacing hnRNP via SMN2 pre-mRNA binding to include exon 7 and result in the full-length SMN RNA and protein to stem the disease and augment motor neurons in spinal muscular atrophy (SMA); (B) Patisiran (Onpattro)-the siRNA binds to the 3′ UTR of the transthyretin (TTR)mRNA facilitating RISC formation and Ago2-cleavage to lower TTR protein production and diminishing amyloid fibrils associated with hereditary transthyretin-mediated (hATTR) amyloidosis. (Bajan and Hutvagner, 2020: © 2020 by the authors. Licensee MDPI, Basel, Switzerland. *This article is an open access article distributed under the terms and conditions of the Creative Commons Attribution (CC BY) license: http://creativecommons. org/licenses/by/4.0/*):

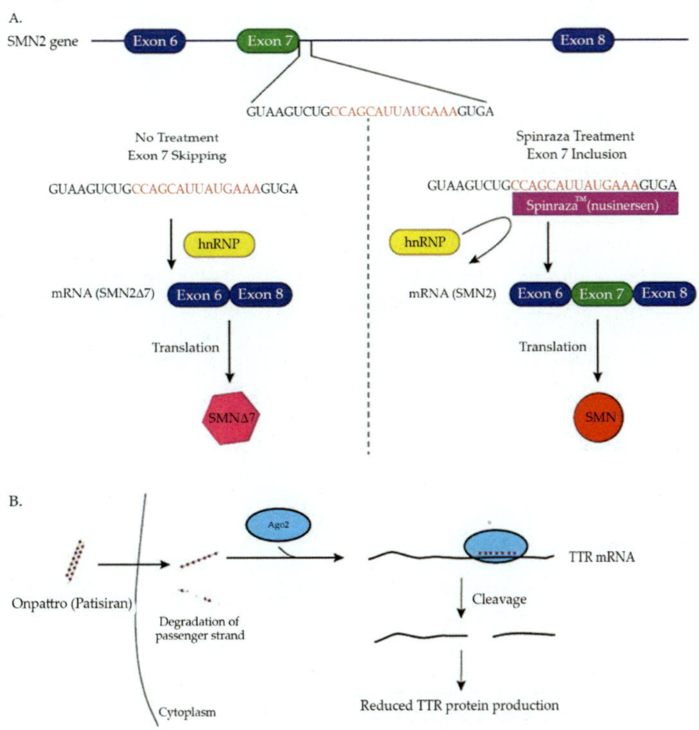

The aforementioned examples along with ongoing research and new technologies such as aptamers, nanomaterials, messenger RNA (mRNA)-based therapeutics can further augment this field. The promise of this approach is highlighted by the FDA approving RNA-based vaccines as the first for emergency use from other candidates! One is BNT162b1 (secreted trimerized receptor binding domain of spike protein) and the other is BNT162b2 (full length spike protein with two proline mutations for augmented prefusion conformation stability): both contain 1-methyl-pseudouridine to augment *in vivo* translation and are lipid nanoparticles. Another example is Moderna RNA vaccine with safety and efficiency on the same lines as that of BNT162b2. Further elucidation of RNA therapeutics can unleash its potential in altering "personalized" treatment (Damase, Sukhovershin, Boada, Taraballi, Pettigrew and Cooke, 2021).

"An element of human invention transforms the natural into the cultural, rendering a higher life form an invention, an effect of the operation of biopower."

— Neil Gerlach, Becoming Biosubjects: Bodies. Systems. Technologies

5.5 REFERENCES

1. Berraondo, P., Sanmamed, M.F., Ochoa, M.C. *et al.* Cytokines in clinical cancer immunotherapy. *Br J Cancer* 120, 6–15 (2019). https://doi.org/10.1038/s41416-018-0328-y
2. Frazzini, S., Riva, F., & Amadori, M. (2021). Therapeutic and Prophylactic Use of Oral, Low-Dose IFNs in Species of Veterinary Interest: Back to the Future. *Veterinary sciences*, *8*(6), 109. https://doi.org/10.3390/vetsci8060109
3. Chulpanova DS, Kitaeva KV, Green AR, Rizvanov AA and Solovyeva VV (2020) Molecular Aspects and Future Perspectives of Cytokine-Based Anti-cancer Immunotherapy. *Front. Cell Dev. Biol.* 8:402. doi: 10.3389/fcell.2020.00402
4. Dickel, S., Grimm, C., Amschler, K., Schnitzler, S. U., Schanz, J., Moerer, O., Payen, D., Tampe, B., & Winkler, M. S. (2021). Case Report: Interferon-γ Restores Monocytic Human Leukocyte Antigen Receptor (mHLA-DR) in Severe COVID-19 With Acquired Immunosuppression Syndrome. *Frontiers in immunology*, *12*, 645124. https://doi.org/10.3389/fimmu.2021.645124
5. Ren X, Zhao M, Lash B, Martino MM and Julier Z (2020) Growth Factor Engineering Strategies for Regenerative Medicine Applications. *Front. Bioeng. Biotechnol.* 7:469. doi: 10.3389/fbioe.2019.00469
6. Koike, Y., Yozaki, M., Utani, A. *et al.* Fibroblast growth factor 2 accelerates the epithelial–mesenchymal transition in keratinocytes during wound healing process. *Sci Rep* 10, 18545 (2020). https://doi.org/10.1038/s41598-020-75584-7
7. Bui, T. Q., Bui, Q., Németh, D., Hegyi, P., Szakács, Z., Rumbus, Z., Tóth, B., Emri, G., Párniczky, A., Sarlós, P., & Varga, O. (2019). Epidermal Growth Factor is Effective in the Treatment of Diabetic Foot Ulcers: Meta-Analysis and Systematic Review. *International journal of environmental research and public health*, *16*(14), 2584. https://doi.org/10.3390/ijerph16142584
8. Bao, G., Tang, M., Zhao, J. et al. Nanobody: a promising toolkit for molecular imaging and disease therapy. EJNMMI Res 11, 6 (2021). https://doi.org/10.1186/s13550-021-00750-5
9. Koenig, P. A., Das, H., Liu, H., Kümmerer, B. M., Gohr, F. N., Jenster, L. M., Schiffelers, L., Tesfamariam, Y. M., Uchima, M., Wuerth, J. D., Gatterdam, K., Ruetalo, N., Christensen, M. H., Fandrey, C. I.,

Normann, S., Tödtmann, J., Pritzl, S., Hanke, L., Boos, J., Yuan, M., … Schmidt, F. I. (2021). Structure-guided multivalent nanobodies block SARS-CoV-2 infection and suppress mutational escape. *Science (New York, N.Y.)*, *371*(6530), eabe6230. https://doi.org/10.1126/science.abe6230

10. Xu, J., Xu, K., Jung, S. *et al.* Nanobodies from camelid mice and llamas neutralize SARS-CoV-2 variants. *Nature* 595, 278–282 (2021). https://doi.org/10.1038/s41586-021-03676-z
11. Damase TR, Sukhovershin R, Boada C, Taraballi F, Pettigrew RI and Cooke JP (2021) The Limitless Future of RNA Therapeutics. *Front. Bioeng. Biotechnol.* 9:628137. doi: 10.3389/fbioe.2021.628137
12. Bajan, S., & Hutvagner, G. (2020). RNA-Based Therapeutics: From Antisense Oligonucleotides to miRNAs. *Cells*, *9*(1), 137. https://doi.org/10.3390/cells9010137
13. Kyriakidis, N.C., López-Cortés, A., González, E.V. *et al.* SARS-CoV-2 vaccines strategies: a comprehensive review of phase 3 candidates. *npj Vaccines* 6, 28 (2021). https://doi.org/10.1038/s41541-021-00292-w

INDEX

D

E

F

G

H

I

K

L

M

N

O

P

R